Preparing for the Third Decade of the National Water-Quality Assessment Program

Committee on Preparing for the Third Decade (Cycle 3) of the
National Water-Quality Assessment (NAWQA) Program

Water Science and Technology Board

Division on Earth and Life Studies

NATIONAL RESEARCH COUNCIL
OF THE NATIONAL ACADEMIES

THE NATIONAL ACADEMIES PRESS
Washington, D.C.
www.nap.edu

THE NATIONAL ACADEMIES PRESS   500 Fifth Street, NW   Washington, DC 20001

NOTICE: The project that is the subject of this report was approved by the Governing Board of the National Research Council, whose members are drawn from the councils of the National Academy of Sciences, the National Academy of Engineering, and the Institute of Medicine. The members of the panel responsible for the report were chosen for their special competences and with regard for appropriate balance.

This study was supported by Grant Number 07HQAG0124 between the National Academy of Sciences and the U.S. Geological Survey. Any opinions, findings, conclusions, or recommendations expressed in this publication are those of the author(s) and do not necessarily reflect the views of the organizations or agencies that provided support for the project.

Cover: Design by Anne Rogers. Map shows total nitrogen yields in kilograms per square kilometer per year determined by the SPAtially Referenced Regressions On Watershed Attributes (SPARROW) model.

International Standard Book Number-13:   978-0-309-26185-2
International Standard Book Number-10:   0-309-26185-6

Additional copies of this report are available for sale from the National Academies Press, 500 Fifth Street, NW, Keck 360, Washington, DC 20001; (800) 624-6242 or (202) 334-3313; http://www.nap.edu.

THE NATIONAL ACADEMIES
Advisers to the Nation on Science, Engineering, and Medicine

The **National Academy of Sciences** is a private, nonprofit, self-perpetuating society of distinguished scholars engaged in scientific and engineering research, dedicated to the furtherance of science and technology and to their use for the general welfare. Upon the authority of the charter granted to it by the Congress in 1863, the Academy has a mandate that requires it to advise the federal government on scientific and technical matters. Dr. Ralph J. Cicerone is president of the National Academy of Sciences.

The **National Academy of Engineering** was established in 1964, under the charter of the National Academy of Sciences, as a parallel organization of outstanding engineers. It is autonomous in its administration and in the selection of its members, sharing with the National Academy of Sciences the responsibility for advising the federal government. The National Academy of Engineering also sponsors engineering programs aimed at meeting national needs, encourages education and research, and recognizes the superior achievements of engineers. Dr. Charles M. Vest is president of the National Academy of Engineering.

The **Institute of Medicine** was established in 1970 by the National Academy of Sciences to secure the services of eminent members of appropriate professions in the examination of policy matters pertaining to the health of the public. The Institute acts under the responsibility given to the National Academy of Sciences by its congressional charter to be an adviser to the federal government and, upon its own initiative, to identify issues of medical care, research, and education. Dr. Harvey V. Fineberg is president of the Institute of Medicine.

The **National Research Council** was organized by the National Academy of Sciences in 1916 to associate the broad community of science and technology with the Academy's purposes of furthering knowledge and advising the federal government. Functioning in accordance with general policies determined by the Academy, the Council has become the principal operating agency of both the National Academy of Sciences and the National Academy of Engineering in providing services to the government, the public, and the scientific and engineering communities. The Council is administered jointly by both Academies and the Institute of Medicine. Dr. Ralph J. Cicerone and Dr. Charles M. Vest are chair and vice chair, respectively, of the National Research Council.

www.nationalacademies.org

Dedication

Walter R. Lynn (1928-2011)

This report is dedicated to Dr. Walter R. Lynn, who served on the committee that authored this report until June 6, 2011, when he passed away.

Dr. Lynn was a member of the faculty of Cornell University for 49 years, serving in various positions, including professor, program director, dean, and university ombudsman. He also served as Mayor of the Village of Cayuga Heights from 2002 to 2008. As a pioneer in the field of environmental systems engineering, he saw the big picture of how water resources and sanitation relate to science, technology, and society. Dr. Lynn was a true interdisciplinarian, and in many ways ahead of his time. For example, he used the term "sustainability" decades ago and interacted frequently with professionals in the medical community on epidemiological matters.

Dr. Lynn was a well-traveled adviser to many organizations. As such, he was a beloved participant in the National Research Council's (NRC's) network of expert volunteers. His service began in 1977 as a member of an august committee charged with assessing future water supply options for the Washington, DC, metropolitan area, and ended with service on the committee that authored the following report. In between, he served on 15 other committees, including several that advised the U.S. Geological Survey and the National Water Quality Assessment program. A National Associate of the National Academies, Dr. Lynn was a favorite of the NRC staff, with a well-deserved reputation as thorough, reasonable, and a pleasure to work with.

Perhaps his most well-known NRC contribution was as founding chair of the Water Science and Technology Board (WSTB) in 1982. In this capacity (1982-1985), he worked tirelessly and thoughtfully with staff in designing important studies and creating from scratch most of WSTB's operating traditions that have endured to the present.

Upon Dr. Lynn's passing, Cornell President David Skorton said of him, "Those who met Walter during his 49 years at Cornell will remember a man of great humor with the exceptional ability to listen and dispense sound wisdom." That is exactly how he will be remembered at the NRC and the WSTB: no pushover, Walter had exceptional skills of modestly imparting sage and thoughtful advice in a style that would cause its recipients to consider and act upon it. The quintessential gentleman-scholar, Dr. Lynn's spirit and memory will continue as a role model for many in the WSTB community, especially the staff who admired and loved working with him.

Preface

After the passage of the Clean Water Act of 1972, state and federal regulatory agencies recognized the inadequacy of the nation's water-quality measurements necessary to assess and address widely recognized water contamination at a national scale. Existing data lacked consistency in their means of collection, methods of analysis, and constituents that were measured. By the middle of the 1980s, Congress, federal and state agencies, and industry collectively understood that the nation needed a comprehensive approach to track and assess water quality and to determine if the quality of water across the nation improved or continued to degrade.

The U.S. Geological Survey (USGS) developed the National Water-Quality Assessment (NAWQA) program to address this need, first with a pilot program in 1986, and then with a widely respected, now mature, national monitoring program. The primary objectives of NAWQA are to assess the status of the nation's groundwater and surface-water resources; evaluate trends in water quality over time; and understand how and to what degree natural and anthropogenic activities affect water quality. Through this three-pronged approach, the NAWQA program provides a national synthesis of the interaction between the natural factors, human activities, and water-quality conditions that affect national water resources.

The first decade (Cycle 1, 1991-2001) of the NAWQA program focused on baseline assessment of the status of water quality in 51 study units that representatively covered approximately two-thirds of the nation's waters. Baseline water quality, in this context, refers to concentrations of measured parameters obtained at the initial samplings. These data would later be compared to future measured values that could document changes in water chemistry caused by long-term effects of regulatory controls over contamination, changing climate, and changing landscape uses. NAWQA in its second decade (Cycle 2, 2001 to the present) built on Cycle 1 through continued monitoring of the study units, but programmatically shifted focus to develop an assessment of observed water quality trends, through, for example, national syntheses on selected water quality parameters and regional assessments of water quality that crossed watershed and political boundaries.

Now, USGS scientists are planning for the NAWQA program's third decade of water-quality assessment (Cycle 3, 2013-2023). They approached the National Research Council's (NRC's) Water Science and Technology Board (WSTB) for perspective on past accomplishments and advice on the current and future design and scope of the program. The committee's task was to assess NAWQA's general accomplishments in Cycle 2 through discussions with program stakeholders and by review of NAWQA's information and products. More importantly, the committee was asked to provide advice on how NAWQA should approach current and future water-quality issues confronting the nation over the next 10 years. Specifically, the committee addressed: (1) present and future water quality issues that should be considered for addition to NAWQA's scope, (2) which NAWQA components should be retained or enhanced, (3) opportunities for NAWQA to better collaborate with others to meet program objectives, (4) the technical soundness of strategic science and design plans for Cycle 3, and (5) NAWQA's ability to meet Cycle 3 objectives.

This request is timely, given the uncertain fiscal climate for the USGS and other governmental agencies. When this NRC committee was being convened, and indeed, during the first 2 years of the review, NAWQA was formulating two strategic documents to guide the program through Cycle 3, the Science Framework and the Science Plan. Captured in these documents are NAWQA's goals for the next decade, namely to move beyond its early focus on nationwide monitoring (Cycle 1) and characterization of water-quality trends (Cycle 2) toward an emphasis on understanding (Cycle 3), the "why and how" of water-quality status and trends. The goal of Cycle 3 is consistent with the original intent of the NAWQA program. To this end, the committee provided two letter reports to help guide NAWQA in shaping its future program.

The committee members brought a wide range of water resources expertise and experience, interacting with NAWQA to make the recommendations herein. Some committee members have provided reviews of NAWQA since its inception through service on earlier NRC committees; other members were users and consumers of NAWQA data and reports. The committee held six deliberative meetings; at the majority of these meetings the committee heard presentations from, and engaged "in discussions with program scientists and others such as users of NAWQA products," as required in the statement of task. Committee members spoke with NAWQA staff; other USGS (non-NAWQA) personnel; local, state, and federal agency "users" of NAWQA data and information; and other users. Committee members attended National Liaison Committee[1] meetings to further understand the needs and role of program stakeholders. The committee also collectively reviewed scores of NAWQA-related reports, both as users and to support this NRC review.

The committee extends thanks to the numerous people external to the USGS who provided highly informative and useful presentations regarding their collective experiences with NAWQA. The committee thanks the USGS NAWQA staff as a whole, particularly Gary L. Rowe and the NAWQA Cycle 3 Planning Team, for answering the many inquiries and requests for reports and documents (Appendix D). The committee also thanks the NRC WSTB staff for their support and leadership.

This report has been reviewed in draft form by individuals chosen for their diverse perspectives and technical expertise in accordance with procedures approved by the NRC's Report Review Committee. The purpose of this independent review is to provide candid and critical comments that will assist the institution in making its published report as sound as possible and to ensure that the report meets institutional standards for objectivity, evidence, and responsiveness to the study charge. The review comments and draft manuscript remain confidential to protect the integrity of the deliberative process. We wish to thank the following individuals for their review of this report: Kenneth R. Bradbury, University of Wisconsin-Madison; David A. Dzombak, Carnegie Mellon University; Jerome B. Gilbert, Consultant, J. Gilbert, Inc.; Ben Grumbles, Clean Water America Alliance; John Melack, University of California, Santa Barbara; Timothy L. Miller, USGS; Karl Rockne, University of Illinois-Chicago; Thomas Theis, IBM Thomas J. Watson Research Center; and Marylynn Yates, University of California, Riverside.

Although the reviewers listed above have provided many constructive comments and suggestions, they were not asked to endorse the conclusions or recommendations nor did they see the final draft of the report before its release. The review of this report was overseen by Henry J. Vaux Jr., University of California. Appointed by the NRC, he was responsible for making certain that an independent examination of this report was carried out in accordance with institutional procedures and that all review comments were carefully considered. Responsibility for the final content of this report rests entirely with the authoring committee and the institution.

This report is intended to assist NAWQA as it enters its third decade

---

[1] See http://acwi.gov/nawqa/, and Chapter 5 and Appendix C of this report.

of nationwide water-quality monitoring and assessment. The committee recognizes that NAWQA is continually striving to improve its efficiency, visibility, and, above all, utility; the committee strongly supports and encourages NAWQA's approach to continuous improvement. It is important that scientists, policy makers, and legislative leaders recognize that identifying and truly understanding water quality status and trends is a long-term undertaking, requiring sustained, long-term support.

Finally, we wish to dedicate this review to committee member Dr. Walter Lynn, who sadly died of cancer during deliberations. Dr. Lynn was a founding member of the WSTB, and he served Cornell University and the water science committee for 49 years in many capacities. He provided the committee with keen insight on the history of the USGS and NAWQA programs, as well as our role in the reviewing process. Cornell President David Skorton called Walter Lynn "one of the most beloved members of the Cornell family," and I can say that all committee members felt the same way about him as a friend and NRC colleague. Walter Lynn will be missed on many levels.

Donald I. Siegel
Chair, Committee to Review the USGS National Water Quality
Assessment (NAWQA) Program

Contents

xiii

Summary

For decades, the U.S. Geological Survey (USGS) has been the primary federal entity responsible for scientific understanding of the nation's surface water and groundwater. As part of this effort, the National Water-Quality Assessment (NAWQA) program assesses the historical and current water-quality conditions and future water-quality scenarios in representative river basins and aquifers across the country. The program was implemented in 1991, primarily in recognition of the importance of understanding the nation's water quality and in response to the conclusion by USGS scientists that their ability to provide information about the nation's water quality at that time was limited. NAWQA objectives are achieved through a design that stresses long-term, standardized collection and interpretation of physical, chemical, and biological data. Water-quality data collection and assessments in river basins and aquifers coupled with regional and national syntheses are the hallmark of the NAWQA program.

Now, the USGS is planning for the third decade of water-quality assessment (Cycle 3, 2013-2023) and approached the National Research Council's (NRC's) Water Science and Technology Board (WSTB) for perspective on past accomplishments as well as current and future design and scope of the program. The NRC responded by forming the Committee on Preparing for the Third Decade (Cycle 3) of the National Water Quality Assessment (NAWQA) Program, appointed under the auspices of the NRC's standing Committee on USGS Water Resource Research. The committee's charge, as laid out in the statement of task, calls for a review of both past accomplishments of the NAWQA program as well as recommendations to improve the design and scientific scope of the program as it moves into its third decade of water-quality assessments. (For the full statement of task, see Box 1-2 in Chapter 1.)

Once the study was under way, the USGS NAWQA Cycle 3 Planning Team asked the committee to give priority to the portion of the task asking for input on scientific priorities for the third decade (Cycle 3) of the NAWQA program. These scientific priorities were expressed in two USGS planning documents, the *Design of Cycle 3 of the National Water Quality Assessment Program, 2013–2023: Part 1: Framework of Water-Quality Issues and Potential Approaches* or the "Science Framework" and the *Design of Cycle 3 of the National Water Quality Assessment Program, 2013-2023: Part 2: Science Plan for Improved Water-Quality Information and Management* or the "Science Plan." The committee responded with two letter reports (Appendixes A and B).[1] This report, the committee's final report, expands upon the advice in the letter reports and addresses the statement of task in its entirety. The report reflects on NAWQA's history and accomplishments (Chapters 2 and 3), outlines a way forward for the program that includes additional feedback on scientific priorities and the Science Plan (Chapter 4), and links this to cooperative, collaborative, and coordinated efforts in the future (Chapter 5).

HISTORY AND ACCOMPLISHMENT OF THE NAWQA PROGRAM

The first decade (Cycle 1, 1991-2001) of the NAWQA program focused on a baseline assessment, i.e., the status of the nation's water-quality conditions. The original program design provided information on water resources by investigating and comparing hydrologically meaningful pieces of geography or study units across the nation. The second decade (Cycle 2, 2001 to the present) focused on identifying trends in water quality, building on the Cycle 1 status activities. During Cycle 2, the program enhanced modeling efforts to extrapolate water quality conditions across the country and expanded communication efforts to disseminate products. In 2004, the program shifted away from the study unit design, restructuring the program design around 8 Major River Basins and 19 Principal Aquifers. This transition is explained in part by the increased emphasis on trend work in Cycle 2 but also by funding decline. This transition is consistent with an overall decline in the number of monitoring sites since 1991 because of planned changes in the design and funding decline (Table S-1).

---

[1] Both the Science Plan and the Science Framework evolved throughout the committee process, responding to continued development from NAWQA leadership, input from stakeholders, and advice from this committee. In the first letter report the committee reviewed the Science Framework version from the fall of 2009. In its second letter report the committee reviewed the Science Plan version from November 2010. The Science Framework is available at http://pubs.usgs.gov/of/2009/1296.

TABLE S-1 The Evolution of NAWQA Program Status and Trends Networks

	Cycle 1		Cycle 2	
	1991-2001	2002-2004	2004-2007	2007-2012
Number of surface water sampling sites	505	145	84	113
Number of aquatic ecology sites[a]	416	125	75	58 (6 sites are ecology only)
Number of groundwater networks[b] and wells	272 networks; 6,307 wells	137 networks; 3,698 wells		

[a] The ecology sites are included in the total number of sampling sites.

[b] A groundwater network is a group of sampling wells.

During these two decades of water-quality monitoring, NAWQA documented that although most water in the United States is fit for many uses, contamination from point and nonpoint sources affected the surface water and groundwater in every study unit, particularly in agricultural and urban areas. Contamination consists of a mixture of nutrients, pesticides, volatile organic compounds, and their breakdown products, which are often just as prevalent as the parent compounds. For example, NAWQA reported that more than half of shallow groundwater samples in urban and agricultural areas contain one or more pesticide compounds. By comparison, pesticides were present in approximately one-third of samples from undeveloped or mixed land use areas. NAWQA also identified improvements in the nation's water quality. For example, after a 2001 federally-mandated phaseout of the organophosphate insecticides diazinon and chlorpyrifos in urban settings, the concentrations of these compounds in northeastern and Midwestern streams decreased after 2002.

NAWQA applied models to support inferences from recent and historical data, project the future water-quality outcome of present and hypothetical actions, and provide the basis for assessing contamination in places where less than optimal or limited field data were available. For example, the SPAtially Referenced Regressions on Watershed attributes (SPARROW) model was used to assess how large-scale changes in land use may affect future nutrient loading from the Mississippi River basin to the Gulf of Mexico. All told, SPARROW models were implemented for six of the eight major U.S. river basins,[2] providing an important resource for assessing water quality at the basin scale and evaluating water management strategies.

In assessing the ecological condition of the nation's surface waters, NAWQA showed that aquatic organisms (algae, macroinvertebrates, and fishes) seldom exhibit similar degrees of alteration in response to different land uses, implying that assessments based on only one type of organism misjudge the extent and severity of impairment. Furthermore, hydrologic alteration and land use change are the major drivers of alterations in ecological condition.

NAWQA distributed and communicated water-quality data through its data warehouse,[3] which makes program data widely available online with sufficient nodes to approximate national coverage and, in some cases, with sufficient regional coverage to assess changes in water quality over time in major watersheds. NAWQA produced approximately 1,900 publications as of January 2012, a publication every 4.2 days on average, a value which, while not an indicator of quality, provides a sense of the quantity of work produced over the history of the program. NAWQA regularly cooperated and coordinated efforts with other programs in the USGS, agencies within the Department of the Interior, and other federal, state, and local agencies. Decision-making, regulatory, and advisory bodies from the federal government (for example, the U.S. Environmental Protection Agency [EPA]), local councils, and state legislatures in more than 30 states used NAWQA's science to the benefit of public health and water resource management. NAWQA studies enabled improvements in areas such as source water protection, quality assurance, quality control, sampling design, sampling methods, analytical protocols, and interpretation frameworks for the water resources issues that states and local governments confront.

The committee concludes that in Cycles 1 and 2, NAWQA provided a successful national assessment of U.S. water quality, in accordance with the mission of a national water-quality assessment program. A more detailed record of representative accomplishments, in no particular order, is presented in Box S-1. NAWQA is well positioned to continue collection and interpretation of water-quality data at a variety of scales, from single rivers and watersheds to larger basins and aquifer systems, and to translate this information to an assessment of the status, trends, and understanding of the nation's water quality. Chapters 2 and 3 of the report reflect further on NAWQA's history and accomplishments.

Despite this record of accomplishment, NAWQA faces many challenges as it moves into Cycle 3 (2013-2023):

---

[2] SPARROW models were implemented for all regions except for California and the Southwest. Models for these regions will be implemented in the future.

[3] See http://water.usgs.gov/nawqa/data.

- How does NAWQA remain a national program in the face of resource decline?
- How should NAWQA balance new status activities against the need to maintain long-term trend networks and understanding studies?
- How can NAWQA use ancillary data[4] and maintain a high level of quality?
- How can NAWQA maintain focus amidst numerous and competing stakeholder demands?

The following sections of this summary correspond to Chapters 4 and 5 of the report, and reflect the committee's advice on a path forward, including specific recommendations in bold.

A WAY FORWARD

The reason for the continuation of the NAWQA program today echoes that which originally motivated the creation of the program: the need to characterize water quality at a national scale. This need persists despite the program's 20-year record of success because of the complex water-related issues facing the nation. Over the past two decades, NAWQA has evolved from a program emphasizing water-quality data collection and trend assessment to one having the potential to forecast contaminant occurrence and aquatic degradation trends under multiple scenarios at nationally significant scales. Although many other successful efforts assess water quality at the local and regional level, NAWQA's unique niche is that it is a *national* program, taking on work that other entities cannot do alone because of, for example, jurisdictional boundaries or available resources. Water-quality monitoring in Cycle 3 is important not only to NAWQA, the USGS, the Department of the Interior, or other agencies, but also to the nation. The federal government needs NAWQA in order to answer the question "Is the nation's water quality getting better or worse?" This is particularly true given that observational networks to measure various water-quality characteristics in the United Status have been on the decline for a number of years. Without measurement, there is no basis on which to evaluate whether policies are effective, no foundation on which to build water management decisions, and no vantage point from which to foresee and forestall water resource challenges. The need for a national water-quality assessment is as important, if not more so today, as when NAWQA was established.

A tipping point for NAWQA is a point where, once crossed, the program as currently organized, scaled, and operated can no longer provide

---

[4] Ancillary data are water-quality data collected by other USGS programs, and national, regional, or local efforts on the same water-quality constituents monitored by NAWQA.

BOX S-1
Accomplishments of the NAWQA Program

National assessment of chemicals in the nation's surface water: NAWQA has provided a national picture of surface water quality.

National assessment of chemicals in the nation's groundwater: This picture extends to the quality of the nation's groundwater, giving the scientific and regulatory communities and the public an understanding of the nation's water quality. Specific to groundwater, NAWQA has demonstrated the utility of groundwater age determination in water-quality studies, especially mixing of old and young waters.

Incorporation of biological indicators of water quality into assessments: NAWQA has integrated measures of indicator organisms into water-quality monitoring and has examined relationships among biological, chemical, hydrological, and land-use parameters using uniform methods at a national scale.

National synthesis reports: These reports synthesize robust data sets using descriptive statistics to draw broad conclusions for the nation to help answer the question that led to the program's development—what is the state of the nation's water-quality?

Continuity and consistency in study methods and design: NAWQA uses standardized sampling regimes, network design, and analytical techniques to enable cross-site comparisons, as well as intensive site-specific and constituent-specific sampling to meet local and regional stakeholder needs, and national water-quality assessments.

---

a national assessment of water quality. Restoration of resources will not reverse this inability to achieve the program's core mission, once the tipping point is crossed. Scaling the program up to what it once was would be inhibited by the break in the long-term monitoring record and the erosion of programmatic infrastructure. However, there may be other scales, modes or organization, and scientific effort that would still allow water-quality monitoring to be achieved. Yet this water-quality monitoring would lack a key feature of the program—national scale—or the ability to say something meaningful about the nation's water quality as a whole.

The committee cannot quantify an exact tipping point for NAWQA. Metrics for identifying when the tipping point is crossed, perhaps built into the network design, would be required. However, the committee can reflect on how to assess proximity to the tipping point through the critical

Development and use of robust extrapolation and inference-based techniques: NAWQA has done an exemplary job of developing and applying robust extrapolation and inference-based models (e.g., SPARROW and the Watershed Regression for Pesticides or WARP models that are statistical, geospatial, and/or process-based and that support inferences from recent and historical data and projections of the outcome of proposed actions).

Information dissemination: NAWQA's communication activities have grown in scope and sophistication as the program has evolved. The program now uses multiple media and appealing graphics to communicate its information products and tools, and it has a wealth of publicly available water-quality data in its data warehouse.

NAWQA science informing policy and management decisions: The program has translated and interpreted its high-quality, nationally consistent data with sophisticated tools so that policy and decision makers can use the program's science to inform efficient decision-making.

Collaboration and cooperation: NAWQA continues to cooperate, coordinate, and collaborate within its own agency as well as with other federal, state, and local agencies in designing and carrying out its programs with a commitment to enhancing its usefulness by making its data and programs relevant to others with interests in water-quality.

Linkages and integration across media, disciplines, and multiple scales: NAWQA has been successful in multidisciplinary research at regional and national scales, collecting and interpreting geographic, hydrologic, biologic, geologic, and climatic data from a range of environmental media (e.g., groundwater, sediments, soils, surface waters, and biota) to help resolve water-quality questions.

question, how much could uncertainty increase in NAWQA outputs before relevant national conclusions could no longer be drawn, and the program suffered irreparable harm? Similarly, does NAWQA have adequate water-quality monitoring data to support its water-quality models?

A successful national water-quality assessment in Cycle 3 would be a national-scale water-quality surveillance program that evaluates and forecasts how changing land use conditions and climate variability may affect water quality in different settings, and that informs water policy and decision makers as they evaluate policy options impacting the nation's water resources. The continuity of national water-quality measurements in space and time is fundamental to this success. **First and foremost, NAWQA's primary focus should be on continuing the monitoring needed to support the national status and trends assessments of the nation's water quality.**

Interruption of the long-term status and trends dataset will limit all other program efforts. Efforts in Cycle 3 that reach beyond the focus of basic monitoring are important (discussed below), but these other goals can only be accomplished if the basic data collection continues.

Measurements provide a snapshot of conditions for only one point in time and are not alone sufficient to forecast future conditions or to understand water quality in unsampled areas. Models are a tool to understanding unsampled areas, constructing scenarios for assessing the impacts of climate and land use change, or forecasting the likely consequences of different policy options. **A focus of NAWQA efforts in Cycle 3, second only to basic monitoring activities, should be the support of NAWQA modeling initiatives.** For example, the committee supports the planned use of the SPARROW model in Cycle 3, expanding the types of contaminants modeled and making the SPARROW model available for public use.

Assessment of the Science Plan

The Science Plan for Cycle 3 is a comprehensive assessment of the nation's needs for understanding status and trends in surface and groundwater quality and developing a portfolio of multiscale models to forecast changes in water quality in response to changes in demographics, land use, and climate. The Science Plan provides a forward-thinking vision for NAWQA science in the next decade of assessing the nation's aquatic resources:

> Science-based strategies can protect and improve water quality for people and ecosystems even as population and threats to water quality continue to grow, demand for water increases, and climate changes.

The Science Plan builds on the existing two decades of data, experience, and NAWQA products. **The overall scope of the Science Plan is broad; thus, the committee recommends that no other issue(s) should be considered for addition to the NAWQA program in Cycle 3.** NAWQA has identified the major water-quality issues facing the nation in the Science Plan.

The Science Plan proposes an expansion of current monitoring networks, similar to the number of sites at the beginning of Cycle 1, and expanding understanding and modeling activities. The Science Plan is structured around four goals, each of which relate to the underlying program principles of status and trends (Goal 1), understanding (Goals 2 and 3), and modeling (Goal 4). The four goals in the Science Plan are consistent with the guiding vision, and contribute to meeting the vision in a synergistic, interconnected, and balanced manner (although not communicated equally well). Then, the Science Plan lists 20 objectives under the four goals that outline the scientific work planned to achieve each goal (Box S-2).

These 20 specific objectives that are described in the Science Plan are not necessarily equal in their contribution to meeting the central or core principles of the Science Plan or to meeting the overall program mission. These objectives also differ in the effort and resources they will require, the clarity of how they are presented, how well they are justified, and the consequences of pursuing them with higher or lower priority. In an ideal world, the Cycle 3 Science Plan would be implemented in full. All 20 objectives have scientific merit. However, given the current federal fiscal climate and the scale of the Science Plan, full-scale implementation of the Science Plan is unlikely.[5]

As directed by the statement of task and to be sensitive to available funding, the committee considered the relative importance of the different scientific objectives within the Science Plan and in terms of trade-offs that implementing one versus the other would represent. The committee categorized the 20 objectives as "essential," "not essential," and those needing "further justification." An objective is essential if it contributes to, for example, monitoring status and trends of surface and groundwater quality and relevant aquatic ecosystem indicators or modeling capabilities and forecasting consequences of future scenarios.[6] An objective that is not essential provides important benefits to the nation and there would be consequences if it were not accomplished, but it is not essential to NAWQA's achievement of its core mission as a national water-quality program. In some cases, these objectives are being addressed by other entities. The Science Plan does not provide sufficient justification of the value to the nation of any objective that needs "further justification."

Objectives corresponding to basic monitoring (i.e., status and trends assessment) and modeling are essential. Basic monitoring activities are the fundamental underpinnings of all program activities (Goal 1). Studies that contribute to modeling that will enable assessments of future scenarios and to estimate water-quality conditions in unsampled waters are critical (Goal 4). Thus, generally speaking, Goal 1 (Objectives 1a, 1d, 1e, 1f, and 1g) and Goal 4 (Objectives 4a and 4b) of the Science Plan are essential. However, it is important to note that embedded within these essential goals are monitoring activities where the committee advises caution because of limited funding. For example, national-scale sediment monitoring is a valuable scientific pursuit. Yet caution is advised given the magnitude of resources likely required to pursue sediment monitoring at the scale and detail proposed in the Science Plan (part of Objective 1e). Similarly, Objec-

---

[5] This supposition is derived from conversations with NAWQA leadership and a set of fiscal scenarios crafted in the Science Framework. These scenarios estimate low, moderate, and high funding levels (compared to fiscal year 2009 levels) and correlate to activities the program could pursue in Cycle 3.

[6] The term "essential" is further defined in Chapter 4.

BOX S-2
NAWQA Cycle 3 Science Plan Goals and Objectives

Goal 1: Assess the current quality of the Nation's freshwater resources and how water quality is changing over time.

(a) Determine the distributions and trends of contaminants in current and future sources of drinking water from streams, rivers, lakes, and reservoirs.

(b) Determine mercury trends in fish tissue.

(c) Determine the distributions and trends in microbial contaminants in streams and rivers used for recreation.

(d) Determine the distributions and trends of contaminants of concern in aquifers needed for domestic and public supplies of drinking water.

(e) Determine the distributions and trends for contaminants, nutrients, sediment, and streamflow alteration that may degrade stream ecosystems.

(f) Determine contaminant, nutrient, and sediment loads to coastal estuaries and other receiving waters.

(g) Determine trends in biological condition in relation to trends and changes in contaminants, nutrients, sediment, and streamflow alteration.

Goal 2: Evaluate how human activities and natural factors, such as land use and climate change, are affecting the quality of surface water and groundwater.

(a) Determine how hydrologic systems—including water budgets, flow paths, travel times, and streamflow alterations—are affected by land use, water use, climate, and natural factors.

(b) Determine how sources, transport, and fluxes of contaminants, nutrients, and sediment are affected by land use, hydrologic system characteristics, climate, and natural factors.

(c) Determine how nutrient transport through streams and rivers is affected by stream ecosystem processes.

(d) Apply understanding of how land use, climate, and natural factors affect water quality to determine the susceptibility of surface-water and groundwater resources to degradation.

(e) Evaluate how the effectiveness of current and historic management practices and policy is related to hydrologic systems, sources, transport, and transformation processes.

---

tive 1a includes lakes and reservoirs. Again, while scientifically valuable, the committee encourages caution when pursuing an objective that has not been traditionally part of NAWQA's design.

Goals 2 and 3 represent the planned extension of Cycle 3 into "understanding" water-quality status and trends, per the original program design (Cycle 1, status; Cycle 2, assessment; Cycle 3, understanding). Many Objectives in Goals 2 and 3 are considered "essential" (2a, 2b, 2d, and 2e; 3b,

Goal 3: Determine the relative effects, mechanisms of activity, and management implications of multiple stressors in aquatic ecosystems.

(a) Determine the effects of contaminants on degradation of stream ecosystems, which contaminants have the greatest effects in different environmental settings and seasons, and evaluate which measures of contaminant exposure are the most useful for assessing potential effects.

(b) Determine the levels of nutrient enrichment that initiate ecological impairment, what ecological properties are affected, and which environmental indicators best identify the effects of nutrient enrichment on aquatic ecosystems.

(c) Determine how changes to suspended and depositional sediment impair stream ecosystems, which ecological properties are affected, and what measures are most appropriate to identify impairment.

(d) Determine the effects of streamflow alteration on stream ecosystems and the physical and chemical mechanisms by which streamflow alteration causes degradation.

(e) Evaluate the relative influences of multiple stressors on stream ecosystems in different regions that are under varying land uses and management practices.

Goal 4: Predict the effects of human activities, climate change, and management strategies on future water quality and ecosystem condition.

(a) Evaluate the suitability of existing water-quality models and enhance as necessary for predicting the effects of changes in climate and land use on water quality and ecosystem conditions.

(b) Develop decision-support tools for managers, policy makers, and scientists to evaluate the effects of changes in climate and human activities on water quality and ecosystems at watershed, state, regional, and national scales.

(c) Predict the physical and chemical water-quality and ecosystem conditions expected to result from future changes in climate and land use for selected watersheds.

SOURCE: Design of Cycle 3 of the National Water Quality Assessment Program, 2013-2023: Part 2: Science Plan for Improved Water-Quality Information and Management

3c, and 3d) because of their scientific importance but also partly because the scientific activities described in these objectives are intimately linked with one another (i.e., one cannot proceed without the other). Basic status and trends monitoring is critical to the proposed understanding studies. Thus, this assessment should also be considered within the committee's overarching recommendation to, first and foremost, maintain status and trends assessment of water quality (i.e., Goal 1).

The committee questions the role of status and trends of microbial contaminants (Objective 1c) in the core vision for NAWQA and considers this objective "not essential." Assessing the status and trends of microbial contaminants at the scale proposed in the Science Plan is a formidable task. The committee questions whether the program has the capacity to proceed with this objective; this could be a resource-intensive effort, and it is inappropriate to proceed at the expense of core efforts, given limited funding. However, the essence of this goal is a human health issue, the result of which would establish the quality of recreational waters. In addition to the obvious scientific benefits, assessing microbial contaminants can be a highly visible activity for the program, clearly demonstrating program impact. An examination of the costs and benefits of obtaining these data when determining whether to pursue this objective is important; collaborative opportunities exist (for example, states and/or the USGS Energy Minerals and Environmental Health Mission Area).

Objective 2c, intended to determine how nutrient transport through streams and rivers is affected by stream ecosystem processes, is a relatively specific objective. This is an important but not essential objective in the committee's view, in part because of potential collaborative opportunities. The committee also considers Objective 3a, effects of contaminants on stream ecosystems, to be a not essential objective for NAWQA. That streams are subjected to multiple stressors is an issue of national importance, but the level of effort required to adequately address this problem could consume a significant amount of the program's resources. Objective 3e, multiple stressors in different regions, is scientifically worthwhile. Yet the committee is concerned with the proposed scale at which these studies will be conducted and how this scale contributes to a national program. Objective 4c (predictions for specific watersheds) depends on the success of the modeling efforts in Objective 4a but also could depend heavily on partnering efforts. Thus, because of the potential collaborative opportunities, the committee considers this objective "not essential."

Determining mercury trends in fish tissue (Objective 1b) needs "further justification" before implementation in Cycle 3, particularly given the scale proposed in the Science Plan. The Science Plan proposes national status and trends monitoring of mercury in fish tissue, expanded from the regional topical study of mercury in fish tissue in Cycle 2. However, consideration of trade-offs is important when evaluating whether the program should pursue this objective. If NAWQA does not pursue national status and trends monitoring of mercury, then other entities (states, other federal agencies, or academia) might provide data, in some cases significantly more data, than would NAWQA. However, further understanding of water-column chemistry and mercury in stream dynamics is a valuable scientific pursuit. Also, NAWQA's Cycle 2 topical study on mercury in fish tissue received significant public interest. By choosing not to pursue the larger scale status and trends assessment of mercury proposed in Objective 1b, the associated public visibility would not be realized.

Although the Implementation Plan for Cycle 3 was not yet prepared at the time of this review, the Science Plan contained preliminary discussion of how to implement the scientific agenda. The Science Plan proposes increased coverage of the NAWQA sampling network to an extent that is similar to that of the original design, coupled with intensive yearly sampling schedules (as opposed to intensive sampling every 2 to 4 years). Although the sense of the committee is that increasing the sampling network is important, some analysis of what would be gained by different numbers and combinations of sites is important. **NAWQA should determine the number of sampling locations and frequency using a similar process that was used in Cycle 2, adapted to the objectives for Cycle 3, with particular consideration of the certainty required for Cycle 3 modeling efforts.**

Communication and Program Impact

NAWQA has used a wide array of approaches to communicate findings, from press releases to congressional briefings, peer-reviewed publications, and the program website. These efforts are an accomplishment, yet communication challenges and opportunities do exist. For example, using tools to bring water-quality data to the public, such as the data warehouse, is an accomplishment of the NAWQA program. Yet the data warehouse, in the committee's judgment, is not user friendly. Furthermore, ensuring that data interpretation, synthesis, and publication of NAWQA data take place in a timely manner is critical. The committee acknowledges the difficulty of this task given the sheer size of the datasets that NAWQA scientists publish, the intense yet valuable USGS peer-review process, and resource constraints. Timely interpretation, synthesis, and release of NAWQA results is critical. NAWQA data used in these results should continue to be delivered to the public via an improved public database.

NAWQA informally measures success and feedback through monitoring the number of website hits, the number of requests for products at the time of release, attendance at briefings during product launches, and collecting information on media coverage. The website homepage contains a link to a document titled *The National Water-Quality Assessment Program— Science to Policy and Management,* which catalogues how stakeholders use NAWQA information and contains personal testimony from a variety of users about the program. NAWQA has conducted three surveys probing satisfaction of customers with specific products and the program at large.[7] However, this tracking of program impact is sporadic and lacks a structured approach and cataloging system. Ultimately, tracking impact will allow NAWQA to demonstrate significance and the return on the nation's investment. A unified strategy for the timely preparation, release, and subsequent tracking of the impact of NAWQA information and products is needed.

Coordination, Cooperation, and Collaboration

The comprehensive nature of the Science Plan makes it clear that NAWQA is committed to being a cooperative, collaborative, and coordinated federal program. This commitment continues and builds on a history of success in these endeavors within USGS, with the Department of the Interior, and with other federal, state, and local agencies. The Science Plan for Cycle 3 is a plan for addressing national water quality needs that deliberately goes beyond what NAWQA can accomplish, providing a framework for other agencies to identify objectives to be met as part of addressing the nation's water quality issues. Thus, although NAWQA will be a cornerstone to implementing the Science Plan, the plan cannot be fully realized without involvement of other groups and agencies and a focus on real collaborative, financial, and intellectual efforts. This will require an expanded approach to involve potential partners and collaborators directly, when appropriate, in the development of science and implementation work plans, explicitly outlining roles, responsibilities, and accountability. The committee recognizes that these efforts are not as simple as they sound and indeed can be costly and time-consuming with attempts to maintain communications among different parties. Difficulties can often arise from overlap or differences in missions that require management time to reconcile. Keeping these potential costs in mind, there is value in NAWQA's ability to leverage greater resources and expertise from external partners to meet the nation's needs for water-quality assessment and understanding.

NAWQA's scope and success have made it a visible and respected focal point within USGS. During the course of the committee's deliberations, and during the time the draft NAWQA Science Plan was under development, USGS reorganized into six mission areas: Ecosystems; Climate and Land-Use Change; Energy and Minerals, and Environmental Health; Natural Hazards; Core Science Systems; and Water. The realignment also created a new Office of Science Quality and Integrity tasked with monitoring and

---

[7] The first Customer Satisfaction Survey was in 2000, probing the usefulness of a specific report, *The Quality of Our Nation's Waters—Nutrients and Pesticides*, Circular 1291. The second and third surveys were more general in format, and were conducted in 2004 and 2010, respectively.

enhancing the quality of USGS science. Although a separate and distinct mission area, water is also a cross-cutting topic important to other themes. NAWQA data and products can fit within most if not all of these mission areas, and opportunities for collaboration should abound from overlapping interests. **NAWQA leaders should seek further opportunities for cooperation, coordination, and collaboration within the USGS and make a systematic effort to communicate its capabilities and potential value to the relevant programs and offices within the USGS through the Science Plan.**

NAWQA has worked to establish cooperative relationships and coordinated efforts with external partners including other federal agencies and state and local authorities. NAWQA's efforts have become important to other agencies, and these relationships have strengthened NAWQA and USGS as a whole. **NAWQA should maintain its interface with the other federal agencies and stakeholder groups and work toward leveraging collaborative resources to meet the needs of the national Science Plan.** For example, in May 2011, the National Oceanic and Atmospheric Administration, the U.S. Army Corps of Engineers, and USGS announced the signing of a Memorandum of Understanding "to form an innovative partnership to address America's growing water resources challenges." NAWQA data and collaboration have contributed to the continuing efforts of the U.S. Environmental Protection Agency (EPA), one of NAWQA's most critical partners, to meet the goals of the Clean Water Act and provided insight on unregulated chemicals under consideration for addition to the Contaminant Candidate List (CCL).[8] This is an example of working toward real collaborative approaches, as urged in this report.

To meet the national needs outlined in the Cycle 3 Science Plan, NAWQA will need to emphasize collaboration in two modes: as a leader that partners with other USGS and external programs, and as a follower with other federal agencies, state and local governments, and the private sector. As part of this approach NAWQA would need to:

- focus on core mission areas where it has unique capabilities, for the program's own implementation efforts;
- leverage resources with other agencies to achieve more of the objectives of the Cycle 3 Science Plan;
- foster higher levels of involvement and investment by other agencies; and
- help others design their own mission-critical programs to meet identified national objectives of the Cycle 3 Science Plan without NAWQA's direct involvement; and

---

[8] EPA's Office of Ground Water and Drinking Water is charged with developing a list of contaminants every 5 years that may require regulation, the CCL.

- explore incentives, for example, access to NAWQA technical assistance, which will enable more sharing of effort for data collection, analysis, and technological innovation across the entire program.

To operate in this more expansive mode, NAWQA should consider engaging partners and collaborators more directly in the development of mutual science plans, seamless exchanges of data and information, and joint implementation of work plans that identify shared responsibilities and accountability. The Cycle 3 Science Plan is a forward-thinking comprehensive water-quality strategy. Because it was authored during a climate of strained federal resources, this is an opportune time for NAWQA to bring together the federal agencies involved in water-quality monitoring and research and, using the Science Plan as a starting point, to develop a collaborative water-quality strategy for the nation.

1

Introduction

Water quality refers to the suitability of water for particular human and ecosystem uses, and operationally it can be defined by federal and state regulatory agencies that establish the physical, chemical, and biological characteristics to adequately meet a particular use. By any measure, water safe to drink, suitable to sustain agriculture, and available to maintain valued natural ecosystems constitutes a fundamental national need. Water quality naturally changes spatially and temporally across the nation because of different climates, seasonality in weather, sources of dissolved and particulate substances, and variability among local and regional hydrologic and geomorphic settings. As would be expected, in some places natural water quality may be insufficient to provide all desired services. However, where suitable water quality does occur, it logically needs to be maintained for the public good.

Failure to maintain water quality occurs throughout the world. In the United States, excess dissolved nutrients from applications of agricultural fertilizer have caused pervasive anoxia where the Mississippi River discharges to the Gulf of Mexico[1] (Osterman et al., 2006). Similar "dead zones" are found throughout the world in marine environments near large river mouths (Diaz and Rosenberg, 2008; Helly and Levein, 2004). Excess nutrients to Chesapeake Bay's estuarine system caused the near collapse of natural fisheries, which only now seem to be returning because of suitable watershed management practices (Harding et al., 1999; NRC, 2004a, 2011c). Finally, excess nutrients from sugar plantations have also com-

---

[1] See http://ecowatch.ncddc.noaa.gov/hypoxia.

17

promised the water quality of the Everglades ecosystem in Florida (NRC, 2008a). The evaporation of water used to irrigate soils in arid western landscapes has caused salinization of western soils and agricultural yields (Schoups et al., 2005).

Prior pervasive surface-water and groundwater contamination of waters in the United States led to the Clean Water Act of 1972,[2] the Safe Water Drinking Act of 1974,[3] and other legislation dealing with water quality. Fortunately, the nation's community water treatment infrastructure remains robust enough to ensure potable, high-quality tap water from rural areas to cities, despite remaining contamination in some places (Moran et al., 2004, 2005, 2007). But, in rural areas, many shallow aquifers no longer are used for drinking water supplies because of nitrate and bacterial contamination originating from agricultural practices and septic systems (Embrey and Runkle, 2006; Nolan and Hitt, 2006). Furthermore, in the north central and northeastern United States, the accumulation of millions of tons of road salt in the unsaturated soil threatens salinization of surface water and shallow groundwaters (Kaushal et al., 2005). Despite these problems, water quality in the United States remains high compared to many other parts of the world, but maintaining this high water quality for human and ecosystem health and prosperity is critical.

Established in 1879, the U.S. Geological Survey (USGS) has a distinguished history of leadership, serving the nation by providing scientific data to describe and understand Earth systems and unbiased assessments to facilitate management of the nation's natural resources. Hydrologic research and hydrologic data collection and analyses are performed through the USGS Water Mission Area, one of six broad earth science mission areas around which USGS is organized: Energy and Minerals, and Environmental Health; Climate and Land-Use Change; Ecosystems; Natural Hazards; Core Science Systems; and Water.[4] The administrative structure of water-related activities at the USGS has evolved throughout the history of the agency, yet the mission has remained constant: "to provide reliable, impartial, timely information needed to understand the nation's water resources."

Because USGS is a science agency with no regulatory or management responsibilities, the Water Mission Area (along with the entirety of the agency) has been widely recognized as a source of unbiased scientific information and hydrologic data. USGS research, studies, and data are used by other

---

[2] Public Law 92-500, the Federal Water Pollution Control Act, or Clean Water Act, is the principal federal law governing contamination of the nation's waters.

[3] Public Law 93-523, the Safe Drinking Water Act, is the principal federal law intended to ensure safe drinking water for the public and applies to every public water system in the United States.

[4] The Office of Science Quality and Integrity is tasked with improving and monitoring the quality of USGS science conducted by the six mission areas.

federal agencies; state, local, and tribal governments; the private sector; and academia as a basis for a wide range of water resources research and water planning and management decisions, including water infrastructure design and maintenance, flood monitoring and emergency notification, drought monitoring, water rights administration, water-quality management, and other related services. USGS carries out its water resources mission through several individual programs spread throughout the agency (Box 1-1) that cumulatively support the nation's hydrologic data network and provide hydrologic assessments at the national, regional, state, and local scale. One of these is the National Water-Quality Assessment (NAWQA) program.

NAWQA was designed and tested in the late 1980s and was implemented at full scale in 1991 to assess historical and current water quality and future water quality scenarios in representative river basins and aquifers across the country. NAWQA's primary objectives are to assess the **status** of the nation's groundwater and surface-water resources; evaluate **trends** in water quality over time; and **understand** how and to what degree natural and anthropogenic activities affect water quality. Taken together, NAWQA's goal is to provide a national synthesis of the interaction between natural factors, human activities, and water-quality conditions to define factors that affect national water resources.

NAWQA's goals are achieved through a design that stresses long-term, standardized collection and interpretation of physical, chemical, and biological water-quality data. NAWQA is not a research program per se; it uses known tools and understanding of processes to probe relevant water-quality topics. Research conducted by USGS's National Research Program and the Toxic Substances Hydrology Program, for example, helps define NAWQA methodologies and topics for the future, but NAWQA does not employ untested methods for probing water quality. Perennial water-quality data collection and sequential assessments in river basins and aquifers as well as regional and national syntheses are key features of the NAWQA program. These activities not only define the status of and trends in water quality, but they also build an evolving understanding of regional and national water quality achieved through careful analysis and interpretation of these long-term water resource data sets.

NAWQA's first decade (Cycle 1, 1991-2001) focused on a baseline assessment of status of the nation's water-quality conditions. The second decade (Cycle 2, 2002 to the present) focused on a more broad-based water-quality assessment, building on the Cycle 1 status monitoring and identifying trends in water quality. Now, USGS scientists are planning for NAWQA's third decade of water-quality assessment (Cycle 3, 2013-2023) and approached the National Research Council's (NRC's) Water Science and Technology Board (WSTB) for perspective on past accomplishments as well as the current and future design and scope of the program. The NRC

BOX 1-1
Water-Related Programs and Activities
at the U.S. Geological Survey

National Water-Quality Assessment (NAWQA) Program: Long-term assessment of water-quality conditions and trends in river basins and groundwater systems nationwide.

National Streamflow Information Program (NSIP): Collection and dissemination of streamflow information that is essential for meeting federal hydrologic information needs.

Cooperative Water Program (Coop Program): Partnerships between USGS and more than 1,500 state, local, and tribal agencies to provide water resources information.

Toxic Substances Hydrology Program (Toxics Program): Field-based research to understand behavior of toxic substances in the nation's hydrologic environments for development of strategies to clean up and protect water quality.

Ground Water Resources Program: Groundwater data collection and the evaluation of controls on regional aquifer systems due to pumping and other stresses.

National Research Program (NRP): Conduct basic and problem-oriented hydrologic research in support of the USGS mission, including investigations of small watersheds (Water, Energy, Biogeochemical Budgets Program).

Office of International Programs: Hydrologic data collection and analysis in support of the global hydrologic community.

Other Water Quality Activities: Analytical capabilities (National Water Quality Laboratory) and data from major rivers (National Stream Quality Accounting Network), from pristine watersheds (Hydrologic Benchmark Network), and from atmospheric deposition (National Atmospheric Deposition Program).

Hydrologic Instrumentation Facility: Instrument development, testing, calibration, and repair; technical support, training, and equipment supply to support hydrologic field activities

Dissemination of Water Resources Information: Physical and chemical data available through the web from the National Water Information System (NWIS);[a] web-based information by states or subjects.[b]

Climate Variability: Understanding the variations in hydrologic conditions due to atmospheric changes and human activities.

Priority Ecosystem Studies: Integrated investigations in large ecosystems of national interest that are impacted by human activity.

Water Institutes: Support of university-based Water Resources Research Institutes in 54 states and territories through grants.

---

[a] See http://water.usgs.gov/NWIS.
[b] See http://water.usgs.gov.

SOURCE: Modified from NRC, 2009.

responded by forming the ad hoc Committee on USGS's National Water Quality Assessment (NAWQA) Program, appointed under the auspices of the standing Committee on USGS Water Resource Research (CWRR). The ad hoc committee's charge, as laid out in the Statement of Task (Box 1-2), calls for a review of both past accomplishments of the program as well as the design and scope of the program as it moves into its third decade of water-quality assessments (Cycle 3).

This report is one of a series of studies that the NRC's Water Science and Technology Board's CWRR has organized. Through these studies, the CWRR has provided advice to the USGS on water-related issues and programs relevant to USGS and the nation since 1985. Over nearly 27 years the CWRR and related committees have overseen reviews of almost every water-related program and initiative, some on a rotating basis. Earlier studies have concerned the National Streamflow Information Program, the National Water Use Information Program, the National Research Program, and the Water Resources Discipline[5] of the USGS as well as areas of research such as river science, groundwater, hazardous materials in the aquatic environment, hydrologic hazards science, and watershed research.

The CWRR has reviewed NAWQA several times in the past. In fact, NAWQA is one of the most "reviewed" USGS programs at the USGS by the NRC. The first was when NAWQA was an unfunded concept and the then chair of the WSTB, Walter Lynn, endorsed the concept of the program in a letter report to then USGS Director Dallas Peck in October of 1985. The most recent NRC advice to NAWQA was the report *Opportunities to Improve the U.S. Geological Survey National Water Quality Assessment Program,* published in 2002. The current study and report was built upon these and other NAWQA reviews by the NRC:

- *Opportunities to Improve the U.S. Geological Survey National Water Quality Assessment Program* (NRC, 2002);
- *National Water Quality Assessment Program: The Challenge of National Synthesis* (NRC, 1994);
- *A Review of the USGS National Water Quality Assessment Pilot Program* (NRC, 1990);
- *National Water Quality Monitoring and Assessment* (NRC, 1987);
- *Letter Report on a Proposed National Water Quality Assessment Program* (NRC, 1985).

Once the study was underway, the USGS NAWQA Cycle 3 Planning Team asked the committee to give priority to its first task concerning

---

[5] The Water Resources Discipline is a former unit under which USGS water-related programs were organized.

BOX 1-2
Statement of Task

Recommendations for the Third Decade (Cycle 3) of the
National Water-Quality Assessment (NAWQA) Program

The project will provide guidance to the U.S. Geological Survey on the design and scope of the NAWQA program as it enters its third decade of water-quality assessments. The committee will assess accomplishments of the NAWQA program since its inception in 1991 by engaging in discussions with the Cycle 3 Planning Team, program scientists and managers, and external stakeholders and users of NAWQA data and scientific information. The committee will also review USGS internal reports on NAWQA's current design for monitoring, assessments, research, and relevance to key water topics. The main activities of the study committee will be to:

1. Provide guidance on the nature and priorities of current and future water-quality issues that will confront the nation over the next 10-15 years and address the following questions:
    • Which issues are currently being addressed by NAWQA and how might the present design and associated assessments for addressing these issues be improved?
    • Are there issues not currently being substantially addressed by NAWQA that should be considered for addition to the scope of NAWQA?
2. Provide advice on how NAWQA should approach these issues in Cycle 3 with respect to the following questions:
    • What components of the program—Surface Water Status and Trends; Ground-Water Status and Trends; Topical Understanding Studies; National Synthesis—should be retained or enhanced to better address national water-quality issues?
    • What components of the program should change to improve how priority issues are addressed?
    • Are there new program components that should be added to NAWQA to enable the program to better address and analyze national water-quality issues and related public policy issues?
3. Identify and assess opportunities for the NAWQA program to better collaborate with other federal, state, and local government, non-governmental organizations, private industry, and academic stakeholders to assess the nation's current and emerging water quality issues.
4. Review strategic science and implementation plans for Cycle 3 for technical soundness and ability to meet stated objectives.

NAWQA's scientific priorities as expressed in *Design of Cycle 3 of the National Water Quality Assessment Program, 2013–2023: Part 1: Framework of Water-Quality Issues and Potential Approaches* or the "Science Framework."[6] More specifically, the Science Framework set out:

> to outline and describe a framework of water quality issues and priorities for Cycle 3 that reflect the unique capabilities and long term goals of NAWQA, an updated assessment of stakeholder priorities, and an emphasis on identifying potential approaches and partners.

The Science Framework represents the first of two planning documents focused on the Cycle 3 design. Eleven topical water-quality priorities were itemized within the two categories, water-quality drivers (climate change, population growth and land use change, and energy and resource development, etc.) and water-quality stressors (sediment, flow modification, emerging contaminants, etc.).

The committee responded to this request with a *Letter Report Assessing the USGS National Water Quality Assessment Program's Science Framework* (Appendix A) published in January 2010. This letter report urges NAWQA to organize its activities around two overarching drivers (or "causes") that indirectly and directly stress water supplies and related ecosystems around the nation: (1) change in land use due to population and other demographic changes and (2) climate variability and change. Under these two broad drivers, the committee encouraged the program to formulate specific, policy-relevant research questions to address and use these questions to identify its scientific priorities and demonstrate program impact. The letter report encouraged NAWQA to further define and enhance program thrusts to meet the principle of "national scale"; adhere to its original program design of probing water-quality "status, trends, and understanding"; align with the new USGS Six Strategic Science Directions; and concentrate on studies where the program can continue to make a unique and substantial scientific contribution.

In a letter dated December 14, 2010, the USGS Director Marsha McNutt asked the committee to provide additional advice on NAWQA's progress in the Cycle 3 planning process, focusing on a second planning document, *Design of Cycle 3 of the National Water Quality Assessment Program, 2013-2023: Part 2: Science Plan for Improved Water-Quality Information and Management* or the NAWQA Cycle 3 or "Science Plan."[7] The purpose

---

[6] At the time of this review, the Science Framework was a working document, available at http://pubs.usgs.gov/of/2009/1296. In its first letter report the committee reviewed the Science Framework version from the fall of 2009.

[7] In its second letter report the committee reviewed the Science Plan version from November 2010.

and scope of this Science Plan is to describe a science strategy for Cycle 3. It outlines four major goals for Cycle 3, the approaches for monitoring, modeling, and scientific studies, partnerships required to achieve the four major goals, and products and outcomes that result from planned assessment activities. The committee was asked to focus on whether the Science Plan sets forth adequate priorities and direction for the future (Statement of Task items 1 and 4).

The committee responded to this request through a second letter report, *Letter Report Assessing the USGS National Water Quality Assessment Program's Science Plan* (Appendix B), published in January of 2011. The committee concluded that the Science Plan is technically sound and that NAWQA has the scientific capability to achieve its objectives. The committee also noted that the program's scientific investments are maturing with the completion of Cycles 1 and 2, enabling NAWQA to move past the current water-quality monitoring to understanding the dynamics of water-quality changes, and using that understanding to forecast likely future conditions under different scenarios of climate and land use change.

This report addresses the entirety of the Statement of Task, augmenting the two previous letter reports. The following chapters reflect on NAWQA's history and accomplishments (Chapter 2 and Chapter 3) and outline a way forward for the program (Chapter 4), which includes an emphasis on collaborative efforts (Chapter 5).

2

NAWQA: Cycle 1 and Cycle 2

Since its beginning, the U.S. Geological Survey (USGS) has been one of the primary federal agencies responsible for assessing the quantity and quality of the nation's surface water and groundwater. In the early 1980s USGS performed and published an assessment of the nation's water, titled *The National Water Summary 1983—Hydrologic Events and Issues* (USGS, 1984). After the completion of this document and related congressional testimony in the mid-1980s, USGS scientists concluded that their ability to say something meaningful about the quality of the nation's waters was limited. Indeed, the USGS resources to assess national water quality were the National Stream Quality Accounting Network[1] (NASQAN) and the Hydrologic Benchmark Network,[2] which, while nationwide, were sparse and were conducting routine monitoring rather than data analysis. Furthermore, NASQAN and the Hydrologic Benchmark Network reflected water-quality sampling approaches from the early 1970s and 1960s, respectively, and thus did not provide data appropriate to address national water-quality questions of the mid-1980s.

Stimulated by the aforementioned events, the USGS contemplated and envisioned a national water-quality assessment program. Key pieces of this original vision included sampling hydrologeologically meaningful units of study or study units, using multiple scales of investigation to achieve a national picture by piecing together information from the study units, integrated teams of scientists performing the water-quality assessment, a

---

[1] See http://water.usgs.gov/nasqan/.
[2] See http://ny.cf.er.usgs.gov/hbn/.

punctuated rotational sampling design, and assessment using established methods (Box 2-1).

Shortly after the NRC's Water Science and Technology Board (WSTB) endorsed the original concept of the National Water-Quality Assessment (NAWQA) program (Chapter 1), it convened a colloquium in 1986 to articulate the necessary elements for a national water-quality assessment program (NRC, 1987). Colloquium participants endorsed the program concept and also raised new issues for consideration such as whether and how to interface with state regulators, which contaminants would be selected for monitoring, and the need to explore surface water and groundwater interactions. For example, the original study unit concept consisted of 123 separate surface water and groundwater units: 69 surface water-dominated and

---

BOX 2-1
The Original Vision for the NAWQA Program

The USGS vision for NAWQA included selecting study units, or hydrologically meaningful pieces of geography (Winter, 2001), in which to monitor water quality. The study units were building blocks for multiple scales of water quality investigation; they served not only as the base level but also as tools for "scaling up" to the bigger, national picture. Consistency between study units would allow the program to make comparable statements about the nation's water quality.

Data collection and data analysis for the water quality assessment in each study unit were to be done by a team working together in an integrated group. This team of scientists was to make measurements, understand what these measurements meant, and make a statement about water quality in a given study unit. It was thought that sampling and assessment should follow a punctuated, rotational system of study with intense data collection for approximately 3 years followed by a period of analysis and publication, a time of minimal monitoring, and a return to the area to repeat the cycle.

NAWQA was envisioned to be a network for data collection defined by geology, hydrology, and land use, rather than a grid or a random sampling strategy. In this way, NAWQA could capture snapshots of both the entire system and "indicator" sites. The design had a strong prejudice toward collecting data in places where USGS had high-quality streamflow data records, in the belief that surface water-quality data are meaningless without considering flow and long-term history. Finally, use of known tools and understanding of processes to monitor the nation's water quality were critical components of the original vision. NAWQA would not deploy untested methods and approaches for analyzing water quality unless on a limited scale. Rather, research and development of methods in other USGS programs would feed the program's activities and assist the program in achieving the goal of assessing the nation's water quality.

SOURCE: R. M. Hirsch, personal communication, May 13, 2009.

54 groundwater dominated (NRC, 1990). However, as the pilot program progressed, it became apparent to both the National Research Council (NRC) committee and USGS that the separate approach had the potential for missing important surface water-groundwater linkages that could have profound effects on the water quality of both systems. Consequently, the decision was made to consolidate groundwater and surface water study units, although most of the study units were either groundwater or surface water dominated.

USGS was authorized by Congress to establish a pilot program in 1986 with seven pilot study units representing a diversity of hydrologic environments and water-quality conditions, four of which were surface water dominated (the upper Illinois River basin in Illinois, Wisconsin, and Indiana; the Kentucky River basin in Kentucky; the lower Kansas River basin in Kansas and Nebraska; and the Yakima River basin in Washington) and three of which were groundwater dominated (the Delmarva Peninsula in Delaware, Maryland, and Virginia; the Carson basin in Western Nevada and Eastern California; and the Central Oklahoma aquifer in central Oklahoma) (NRC, 1990). USGS requested the NRC to undertake a 2-year evaluation of the pilot studies in 1987, and the NRC responded with *A Review of the USGS National Water Quality Assessment Pilot Program* (NRC, 1990). This NRC committee was invited to assist in the evolution and refinement of the NAWQA design as it moved toward full-scale implementation, deliberating on several NAWQA planning documents, issuing an interim report, and visiting the seven pilot study units. The NRC committee was supportive of the NAWQA effort (Box 2-2).

The success of the pilot effort led to NAWQA's full-scale implementation in 1991 with the program goals of status, trends over time, and understanding as cornerstones of the program mission—cornerstones that have not changed through the evolution of the program. At the time of its

---

BOX 2-2
Perspective from NRC (1990)

"The [NRC] committee is convinced that there is a genuine need for a long-term, large spatial scale national assessment of water quality in the United States. Human health and environmental health are inextricably linked to our nation's water quality. . . . The [NRC] committee is convinced that a national scale, long term water quality assessment is in the best interest of the country."

SOURCE: NRC, 1990.

conception, NAWQA was the largest water resources program ever under-taken by USGS (R. J. Gilliom and R. M. Hirsch, personal communication, May 13, 2009).

CYCLE 1 OVERVIEW

In the first decade of water-quality monitoring (Cycle 1, 1991-2001) NAWQA set out to (1) accumulate high-quality, multidisciplinary, water-quality data and (2) generate a national synthesis of those data focusing on analysis of the highest-priority issues that cuts across the geography and answers the question, "How is the nation's water quality changing?" The program demonstrated considerable progress toward a national water-quality assessment in Cycle 1. For thoroughness and to place this report in context, the committee notes key components of Cycle 1 here. (For a detailed review of Cycle 1 see NRC [2002].)

The Study Unit Concept

The Cycle 1 study units accounted for 60 to 70 percent of the nation's water use and population served by public water supplies and covered about one-half of the land area of the United States. A broad suite of physi-cal, chemical, and biological constituents was selected based on relevance to water-quality issues and existing analytical methods including measure-ments of:

- streamflow,
- pH,
- temperature,
- dissolved oxygen,
- specific conductance,
- major ions,
- nutrients,
- trace elements,
- organic carbon,
- pesticides, and
- volatile organic compounds (VOCs) (NRC, 2002).

Also, descriptions of biological communities were made based on different taxonomic groups and habitat conditions (NRC, 2002). A suite of surface water reference sites, a sampling site selected for relatively undisturbed conditions, was built into the surface water network design. At the end of Cycle 1, monitoring at 51 study units plus a study of the High Plains Aqui-fer in the central United States were completed. (The geographic scope of the original design was 59 study units, which was adjusted to 51 to account for fiscal restrictions.) The High Plains Aquifer study was a pilot study for a regional approach to a groundwater assessment in the southern High Plains and was added near the end of Cycle 1.

In three groups over time, the study units were phased in during Cycle 1: study units 1-20 in 1991, study units 21-36 in 1994, and study units 37-51 in 1997 (Figure 2-1). At the onset, each study unit had a 2-year startup phase with time for planning and analysis of existing data, which was a major effort. At the same time, each study unit was developing liaison committees with local stakeholders, which became critical to guide how each study unit analysis was carried out and how the results were used to enhance water management. Within each study unit, an integrated group of scientists addressed the three primary objectives by (1) making measurements, (2) evaluating these measurements to understand water quality, and (3) making statements about what is learned and known about a particular study unit. After the 2 year startup, each study unit entered a 3 year intensive data-collection stage. This was followed by a period of data analysis and completion of major reports and then low-level monitoring

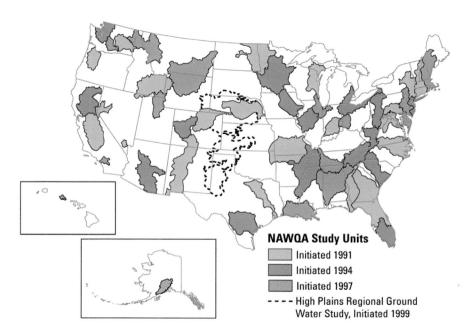

FIGURE 2-1 Cycle 1 study units (51 plus the High Plains Aquifer) SOURCE: R. J. Gilliom, personal communication, May 17, 2010.

and assessment activities. Following a short period of retrospective analysis, each study unit would ramp back up and enter the intensive data-collection phase again—10 years after the previous data-collection phase (Figure 2-2).

This fixed site design with periodic rotational sampling allowed NAWQA to collect data at regular snapshots in time and document trends. Sampling a total of 505 stream sites and more than 6,000 groundwater wells, each study unit assessment resulted in many individual publications. At the end of 2001, more than 1,000 NAWQA publications were available (NRC, 2002). Also, the study units effectively bridged the environmental system because of a tailored sampling strategy in each study unit (groundwater and/or surface water; the water column and/or bed sediment; pesticides and/or nutrients) and a diverse team of scientists working on each assessment. The similar design of each study unit investigation and the use of standard methods made it possible to compare results between different study units, thus enabling multiple scales of investigation or regional and national assessments. These regional and national assessments, referred to as "national syntheses," aggregated water-quality information and also allowed for analysis of important national issues such as, for example, non-point source pollution.

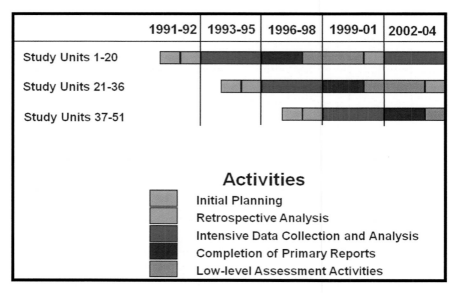

FIGURE 2-2 The phase in and cycling of NAWQA study units. SOURCE: R. J. Gilliom, personal communication, May 17, 2010.

National Synthesis

NAWQA phased in national synthesis assessments during Cycle 1, conducted by national synthesis teams. These included pesticides and nutrients in 1991, VOCs in 1994, and trace elements and ecology in 1997. Criteria for the selection of these topics considered a combination of understanding stakeholder priorities, capturing appropriate scale (i.e., topics should affect a large area or many small areas), representing persistent and recurring issues, importance to the study units that were in place, and complementing other national synthesis topics. NRC (2002) commended NAWQA for its groundbreaking work in these areas during Cycle 1.

Environmental Framework

NAWQA activities were developed with an "environmental framework" or a broader context through which the data were related to the bigger, environmental picture. This framework, composed of "common natural and human-related factors, such as geology and land use," was used "to compare and contrast findings on water quality within and among study units in relation to causative factors and, ultimately, to develop inferences about water quality in areas that have not been sampled" (Gilliom et al., 1995). The environmental framework was reflected in the entire program design from sampling type to the interdisciplinary staffing structure. Application of the environmental framework assisted the program in, for example, choosing a drainage basin to study or a set of indicator sites. The environmental framework concept was and is today a touchstone for program efforts.

CYCLE 2

The second cycle of water-quality monitoring (Cycle 2) began in 2002 and extends to the end of fiscal year (FY) 2012, slightly past the duration of this committee's review. Per the original design, NAWQA implemented a shift toward trends and understanding as the program moved out of Cycle 1. NAWQA integrated a number of new components as a result of evaluations from the Cycle 2 National Implementation Team (NIT), input from NAWQA personnel who were the primary drivers of the original design, and recommendations from the 2002 NRC report.[3] NAWQA investigated select new contaminants and addressed many complexities involved with their environmental occurrence such as seasonal variations, degradation products, and chemical mixtures. These new activities were pursued

---

[3] Approximately 80 percent of the 2002 NRC recommendations were implemented by NAWQA, and those that were not were omitted largely because of funding restrictions.

through program components such as Topical Studies and the Source Water Quality Assessment, discussed in the following pages. However, because of limited funding NAWQA was unable to pursue the following recommendations from the NRC report (2002): sample lakes and reservoirs that are important sources of water supply; enhance sediment monitoring, enhance interpretation, and make sediment a topic of a national synthesis team; and add pharmaceuticals, high production volume chemicals, and waterborne pathogens and microbial indicator organisms to the list of contaminants monitored in Cycle 2. The program also continued to assess the current water quality of the nation through standardized data collection, in concert with the goal of assessing long-term water-quality trends. Planned activities were grouped into 12 themes:[4]

1. resources
2. drinking water sources
3. contaminants
4. trends in status
5. response to agricultural management
6. response to urbanization
7. sources of contaminants
8. transport to and within groundwater
9. transport to and within streams
10. groundwater and surface water interactions
11. effects on aquatic biota
12. extrapolation

Each theme correlated to NAWQA's goal of status (themes 1-3), trends (themes 4-6), and understanding (themes 7-12).

In Cycle 1, NAWQA focused 80 percent of program resources on the status effort, continuing to establish the nation's baseline water-quality condition. This was reduced to 20 percent of available resources in Cycle 2, although NAWQA did enhance the status activities with the Source Water Quality Assessments, an examination of the drinking water in communities across the United States, corresponding to status theme 2 (Delzer and Hamilton, 2007). The program placed an increased emphasis on trends (40 percent of program resources) and understanding (40 percent of program resources) through planned topical studies with a source, fate, and transport perspective (Figure 2-3).

This shift in design at the onset of Cycle 2 along with several years of flat funding required beginning Cycle 2 with 42 study units, instead of the

---

[4] Items 1 and 3 (resources and contaminants) were not pursued in Cycle 2 because of limited funding.

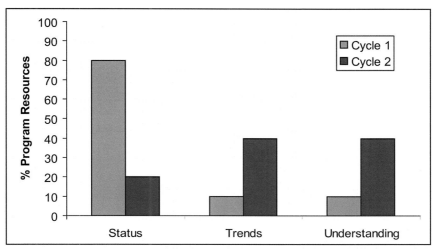

FIGURE 2-3 Shift from status (Cycle 1) to trends and understanding (Cycle 2). SOURCE: G. Rowe, personal communication, May 13, 2009.

51 monitored in Cycle 1 (Figure 2-4). NAWQA conducted a detailed analysis to determine which study units should be discontinued or consolidated and which were the most representative study units. Discontinued study units include those in Hawaii (the Oahu Study Unit), Alaska (the Cook Inlet Basin Study Unit), and the Lower Susquehanna basin in Pennsylvania (the Lower Susquehanna River Basin Study Unit). For example, the decision was made to discontinue the Hawaii study unit because of low population density relative to water use in comparison with other study units. Low population density or low water use criteria drove the discontinuance of most of the other study units as well.

As Cycle 2 progressed, perhaps the most notable design change began in 2004. The program transitioned away from the study unit focus and moved to a larger-scale regional design for status and trends assessment because of limited resources. The regional design retained a core of status and trends monitoring still conducted within the study units, but de-emphasized the role of more detailed study unit investigations and their individual teams and liaison committees. Status and trends data analysis and modeling, as well as program products, were shifted to teams organized by 8 Major River Basins (MRBs) and 19 Principal Aquifers (PAs) (Figures 2-5 and 2-6).

The MRB and PA regions are similar in concept to the role of study units as the building blocks of Cycle 1, but on a larger scale that collectively includes the conterminous United States, albeit at lower resolution. Cor-

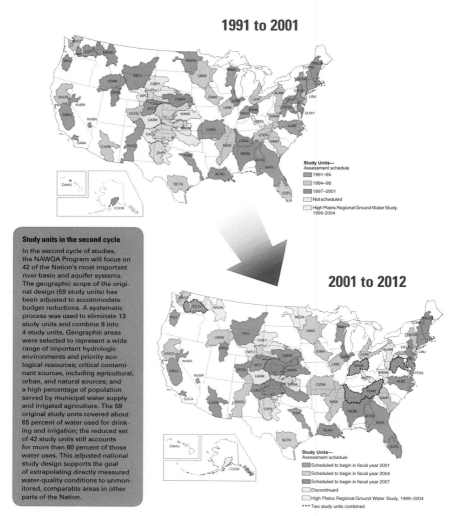

FIGURE 2-4 The planned reduction and consolidation of study units at the onset of Cycle 2. Discontinued study units are shown in yellow. See Gilliom et al. (2001) for study unit designation. SOURCE: Gilliom et al., 2001.

responding to the study unit redesign, monitoring for specific conductance and temperature ceased, and pesticide monitoring at reference sites was discontinued. Also, the role of study unit liaison committees was reduced, which in turn reduced the degree of local stakeholder input to NAWQA (see Chapter 5 for further discussion).

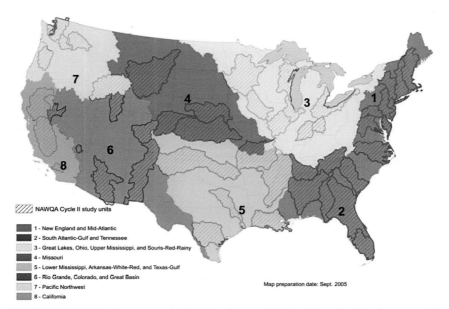

FIGURE 2-5 Eight large geographical regions or Major River Basins that were the basis of NAWQA's status and trends assessment in the latter portion of Cycle 2 (2006-2012). SOURCE: Crawford et al., 2006.

NAWQA expanded efforts toward modeling in Cycle 2, to allow the program to extrapolate water-quality conditions across the country in areas not sampled by the program. This began in 2002 with an assessment of nutrient conditions in six large regions across the country using the SPAtially Referenced Regressions on Watershed Attributes (SPARROW) model (Smith et al., 2003). Later, mid-Cycle 2, the shift from study units to MRBs and PAs was considered an opportune time to begin developing planned regional-scale water-quality models. For example, a regional-scale SPARROW model was developed for the southeastern United States (Hoos and McMahon, 2009).

NAWQA increased efforts to communicate and disseminate its products and information. NAWQA moved from dissemination through paper reports in Cycle 1 to a multimedia in Cycle 2. Communication strategies were created for each major report, and more web-based dissemination and decision-support tools were initiated to reach a variety of audiences. Components of the enhanced communication effort included[5]:

_____

[5] NAWQA leadership, personal communication, May 9, 2009.

- improvements to the NAWQA website,
- creation of a publication search engine,
- multi-level rollouts of high-visibility findings,
- detailed communication plans for visible reports,
- improved data warehouse with data mapping capability,
- web-based decision-support tools, and
- podcasts.

Chapter 3 and Appendix C contain further discussion of the NAWQA communication effort.

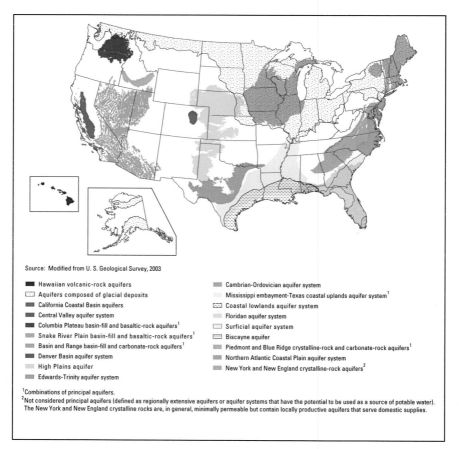

Source: Modified from U. S. Geological Survey, 2003

- ■ Hawaiian volcanic-rock aquifers
- ⬚ Aquifers composed of glacial deposits
- ■ California Coastal Basin aquifers
- ■ Central Valley aquifer system
- ■ Columbia Plateau basin-fill and basaltic-rock aquifers[1]
- ▦ Snake River Plain basin-fill and basaltic-rock aquifers[1]
- ■ Basin and Range basin-fill and carbonate-rock aquifers[1]
- ■ Denver Basin aquifer system
- ▨ High Plains aquifer
- ▦ Edwards-Trinity aquifer system
- ▦ Cambrian-Ordovician aquifer system
- ▨ Mississippi embayment-Texas coastal uplands aquifer system[1]
- ⬚ Coastal lowlands aquifer system
- ▨ Floridan aquifer system
- ⬚ Surficial aquifer system
- ▨ Biscayne aquifer
- ▦ Piedmont and Blue Ridge crystalline-rock and carbonate-rock aquifers[1]
- ▦ Northern Atlantic Coastal Plain aquifer system
- ▦ New York and New England crystalline-rock aquifers[2]

[1]Combinations of principal aquifers.
[2]Not considered principal aquifers (defined as regionally extensive aquifers or aquifer systems that have the potential to be used as a source of potable water). The New York and New England crystalline rocks are, in general, minimally permeable but contain locally productive aquifers that serve domestic supplies.

FIGURE 2-6 Nineteen Principal Aquifers selected for regional assessment during the latter portion of Cycle 2 (2006-2012). SOURCE: Lapham et al., 2005.

Status and Trends Networks

As NAWQA moved from Cycle 1 to Cycle 2 in the midst of planned and unplanned program design changes, the status and trends sampling networks also changed. As described above, these changes emphasized regional assessment and resulted in more regional-scale analysis in Cycle 2.

Surface Water Status and Trends Network

In Cycle 1 505 surface water sites were sampled in 3-year, intensive, water-quality sampling periods per the original design. Of the original 505 stream sites monitored in Cycle 1, 145 were selected for annual trends monitoring at the start of Cycle 2 as specified in the Cycle 2 NIT's report, *Study Unit Design Guidelines for Cycle II of the National Water Quality Assessment Program*[6] (Gilliom et al., 2000). However, by 2005 available funding could only support monitoring of 84 sites annually, which lasted about 2 years until 2007. Since 2007, NAWQA has maintained 113 sites at the expense of an annual sampling strategy; a rotational design was employed where the majority of the sites were sampled 1 year out of every 4 years (Table 2-1 and Figure 2-7). Twelve of these sites, the larger river integrator sites or sites on large rivers that drain significant agricultural and urban areas, are sampled yearly. All Cycle 2 sampling sites were selected from Cycle 1 sites for NAWQA to preserve and maintain a long record of consistent data that is useful for trend analysis (G. Rowe, personal communication, May 17, 2010).

Aquatic Ecology Status and Trends Network

The aquatic ecology sampling network was cut back even more than the surface water network in Cycle 2, to 58 sites, rotated and sampled every other year (Table 2-1). NAWQA divided the country into two sections, western and eastern, with sampling rotating back and forth along with detailed investigations that continue today. NAWQA's philosophy on selecting these sites used environmental framework as a touchstone, with two key components: (1) choose representative sites and (2) scale the sampling design to accommodate the size of the river. During sampling, a detailed habitat assessment is performed—algae, invertebrates, and fish samples are taken at each site along with a riparian assessment—assessing the biological status of each site. All sites are co-located with sites where water chemistry, bed sediment, and streamflow sampling occur. This pro-

---

[6] The NIT's report describes the design and implementation strategy for Cycle 2 investigations. The report *Opportunities to Improve the U.S. Geological Survey Water Quality Assessment Program* reviewed this report (NRC, 2002).

TABLE 2-1 The Evolution of NAWQA Program Status and Trends Networks

	Cycle 1		Cycle 2	
	1991-2001	2002-2004	2004-2007	2007-2012
Number of surface water sampling sites	505	145	84	113
Number of aquatic ecology sites[a]	416	125	75	58 (6 sites are ecology only)
Number of groundwater networks[b] and wells		272 networks; 6,307 wells	137 networks; 3,698 wells	

[a] The ecology sites are included in the total number of sampling sites.
[b] A groundwater network is a cluster of sampling wells.

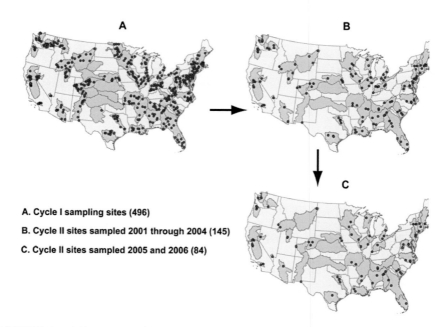

A. Cycle I sampling sites (496)

B. Cycle II sites sampled 2001 through 2004 (145)

C. Cycle II sites sampled 2005 and 2006 (84)

FIGURE 2-7 NAWQA Surface Water Status and Trends Network redesigns in the continental United States from Cycle 1 (A) through Cycle 2 (B and C), due to lower declining resources. Cycle 1 sampling sites shown here do not include nine surface water sites in Alaska and Hawaii. SOURCE: G. Rowe, personal communication, May 9, 2009.

duces a "reach assessment" where NAWQA probes what organisms are exposed to in a given watershed. To enhance NAWQA's ability to use these data to provide a national assessment of ecological conditions given limited sampling, the program collaborated with the U.S. Environmental Protection Agency's (EPA's) Environmental Monitoring and Assessment Program (EMAP)[7] to paint a picture of water quality in the western United States based on indicator organisms (Carlisle and Hawkins 2008). (For additional information, see Chapter 5.)

Groundwater Status and Trends Network

In Cycle 1 NAWQA had approximately 272 groundwater networks or clusters of sampling wells, for a total of 6,307 wells sampled throughout the study units for groundwater status and trends (Figure 2-8, top). In Cycle 2, NAWQA viewed activities on the basis of Principal Aquifers representing a more regional view than Cycle 1, i.e., Principal Aquifer Assessments. The program sampled 137 groundwater networks for a total of 3,698 wells (Table 2-1, Figure 2-8, bottom), evaluating conditions and trends in 19 regional aquifers per the NAWQA Cycle 2 Implementation Team's original design (Gilliom et al., 2000). These 19 aquifers account for approximately 75 percent of the estimated withdrawals of groundwater for drinking water supply in the United States (Lapham et al., 2005).

Unlike the Surface Water Status and Trends Network and the Aquatic Ecology Status and Trends Network, the groundwater network design remained largely unchanged through the duration of Cycle 2. This design included decadal sampling of all wells, biennial sampling of 5 wells within each network, and seasonal sampling of wells selected for biennial sampling (Gilliom et al., 2000). This was, in part, due to the relatively modest scale of the original Cycle 2 Groundwater Status and Trends Network design but also to the slow hydrologic response time of groundwater, permitting more flexibility in correlating the timing of, for example, biennial sampling with budgetary realities (NAWQA leadership, personal communication, July 20, 2012).

Status and Trends Assessments and Activities

During Cycle 2, NAWQA mined the 10 years of monitoring data from Cycle 1, augmented by continued monitoring in Cycle 2, to determine and publish long-term assessments of trends in the nation's water quality. The program synthesized data from the Surface Water Status and Trends Network, Aquatic Ecology Status and Trends Network, and Groundwa-

---

[7] See http://www.epa.gov/emap/.

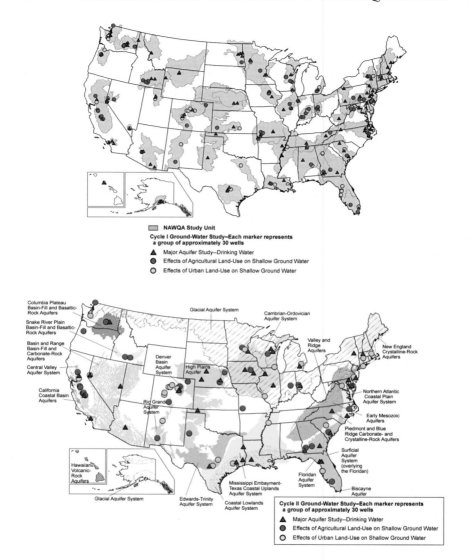

FIGURE 2-8 Cycle 1 (top) and Cycle 2 (bottom) groundwater status and trends networks. Each marker represents a group of sampling wells. SOURCE: USGS leadership team, personal communication, May 9, 2009.

BOX 2-3
Reports on Water-Quality Trends in the Major River Basins

1. *Trends in Nutrient Concentrations, Loads, and Yields in Streams in the Sacramento and Santa Ana Basins, California, 1975-2004* (Kratzer et al., 2011)
2. *Nutrient and Suspended-Sediment Transport and Trends in the Columbia River and Puget Sound Basins, 1993-2003* (Wise et al., 2007)
3. *Trends in Nutrient and Sediment Concentrations and Loads in Major River Basins of the South-Central United States, 1993-2004* (Rebich and Demcheck, 2007)
4. *Nutrient and Suspended-Sediment Trends in the Missouri River Basin, 1993-2003* (Sprague et al., 2007)
5. *Trends in Streamflow, and Nutrient and Suspended Concentrations and Loads in the Upper Mississippi, Ohio, Red, and Great Lakes River Basins, 1975-2004* (Lorenz et al., 2009)
6. *Trends in Pesticide Concentrations in Corn-Belt Streams, 1996-2006* (Sullivan et al., 2009)

---

ter Status and Trends Network and released a variety of products. Many water-quality trends emerged at the local, regional, and national levels. For example, NAWQA showed that the investment by the Bureau of Reclamation in improving the water quality of the Colorado River resulted in a decrease in dissolved solids downstream of salinity control projects (Anning et al., 2007). A select group of status and trends publications and results are highlighted here; Chapter 3 and other areas of this report continue this discussion.

At the time of this report, six trends reports exist for the major river basins on nutrients, sediment, and pesticides (Box 2-3).[8] NAWQA also has published a variety of information assessing the PAs.[9] For example, the High Plains Aquifer Professional Paper summarizes the water quality in this aquifer and was NAWQA's first systematic regional assessment of groundwater (McMahon et al., 2007). Also, NAWQA National Synthesis Teams synthesized the results from NAWQA investigations for priority water-quality issues (pesticides, nutrients, VOCs, ecology, and trace elements) and produced capstone reports on pesticides, nutrients, and VOCs in Cycle 2. The National Synthesis Assessments are discussed further in Chapter 3.

---

[8] For a complete list of publications on MRBs, see http://water.usgs.gov/nawqa/studies/mrb/pubs.html.

[9] For a complete list of publications on PA Assessments, see http://water.usgs.gov/nawqa/studies/praq/.

NAWQA's Source Water Quality Assessments (SWQA) examined drinking water quality of community water systems across the United States by comparing compounds in raw ambient water collected at a supply well or surface-water intake prior to treatment (i.e., "source water") to compounds in the finished water supplied to the community (Delzer and Hamilton, 2007). The assessment focused on 280 unregulated organic compounds with a focus on VOCs and pesticides. Carter et al. (2007) provide information on the design and analytical methods used in the SWQA. While a diverse group of compounds were present in source water, the majority of the compounds assessed were present only at low concentrations (<< 1 ppb). Compounds detected in source water were often in finished water, although compounds detected in finished water were below human-health benchmarks if one existed. Mixtures of compounds were commonly detected in both. Capstone products were released in 2008 and 2009 (Hopple et al., 2009; Kingsbury et al., 2008).

Understanding Activities

The understanding component of NAWQA was carried out in Cycle 2 through five hypothesis-driven topical studies. The conceptual approach of these studies was to understand contaminant source, fate and transport, and impacts on humans and aquatic ecosystems. NAWQA took a mass balance approach to the studies, understanding that a mass balance of water and a mass balance of constituents go hand in hand (i.e., scientists should understand how water is flowing through the system in order to eventually understand the effects of contaminants). NAWQA integrated the use of models into a few of the topical studies. With each topical study, NAWQA adhered to the concept of a national program with a focus on a national understanding of water-quality problems. In each of the five topical studies, NAWQA probed multiple locations, scales, and gradients (i.e., multiple climate, landscape settings, hydrology, crops, land use settings, and atmospheric deposition settings). The topical studies were nested within the study units of Cycle 1, using knowledge gained in Cycle 1:

- Topical Study 1: Agrochemical Sources, Transport, and Fate[10]
- Topical Study 2: Effects of Nutrient Enrichment in Stream Ecosystems[11]
- Topical Study 3: Mercury Cycling in Stream Ecosystems[12]

---

[10] See http://pubs.usgs.gov/fs/2004/3098/.
[11] See http://wa.water.usgs.gov/neet/.
[12] See http://water.usgs.gov/nawqa/mercury/.

- Topical Study 4: Effects of Urbanization on Stream Ecosystems[13]
- Topical Study 5: Contaminant Transport and Public Supply Wells[14]

The topical studies produced a variety of interesting findings, published in methods papers, comprehensive journal article series, and USGS reports. Due, in part, to an underestimation of the amount of work associated with these efforts, some topical studies progressed further than others during Cycle 2. For example, the mercury study (Topical Study 3) documented methylmercury concentrations across the United States and observed that the highest levels of methylmercury in fish are found in the southeastern United States and in mined areas in the western United States (Scudder et al., 2009). (Methylmercury is the most toxic form of mercury in the environment and is readily taken up by aquatic organisms.) NAWQA noted that major urban centers are experiencing a significant increase in mercury deposition. Finally, because of biogeochemical properties of methylmercury, concentrations of the contaminant in streams are driven by wetland density and dissolved organic carbon concentration (Figure 2-9).

Monitoring . . . to Monitoring and Modeling . . . to the User

In Cycle 2, NAWQA moved from monitoring to monitoring and modeling water quality of the nation's groundwater, surface water, and ecology at all scales (i.e., using deterministic models at smaller scales and statistical regression at large scales). The NAWQA Cycle 2 modeling approach is to use monitoring data and stream network to probe water quality from the regional and national to the local scales. Modeling efforts amplify the program goals through (1) extrapolation of water-quality conditions to unmonitored areas to facilitate a "national assessment" and (2) forecasting of conditions and simulation of the effects of changes in influencing factors (test scenarios). As Cycle 2 draws to a close, the modeling efforts are improving understanding of the factors (sources, transport, etc.) that influence water quality.

The goal of one of NAWQA's first exercises in modeling was to predict groundwater vulnerability to nitrate contamination at the national scale. The program showed this vulnerability based on monitoring data, fertilizer data, and soil characteristics, which were integrated into a model called GWAVA (Ground-WAter Vulnerability Assessment). In the southeastern United States NAWQA reported lower concentrations of nitrogen where denitrification is promoted compared to the central plains (Nebraska), where the United States has high fertilizer use, gravel and sand, fast transport, and

---

[13] See http://water.usgs.gov/nawqa/urban/.
[14] See http://oh.water.usgs.gov/tanc/NAWQATANC.htm.

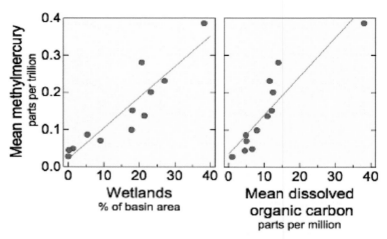

FIGURE 2-9 Increasing concentrations of mean methylmercury in U.S. streams with increased wetland density and mean dissolved organic carbon. Once deposited in wetlands, mercury is converted to methylmercury. Dissolved organic carbon binds strongly to mercury keeping mercury in the aquatic zone and available for uptake by organisms. SOURCE: USGS, 2009a.

lack of denitrification (Nolan and Hitt, 2006). EPA uses this information to help prioritize monitoring and better assess its regulatory efforts.

During Cycle 2 NAWQA developed empirical models to probe hydrologic alteration nationwide as well as the connection between hydrologic alteration and the structure of macroinvertebrates and fish assemblages. NAWQA successfully modeled ecologically important flow metrics under a "natural" or "minimally disturbed" flow regime using geospatial data and a reference condition approach. This opened the possibility of quantification of hydrologic alteration across the United States (Figure 2-10). Using geospatial models and NAWQA data, Carlisle et al. (2011) demonstrated that diminished magnitude of flows was the best predictor of impairment of macroinvertebrate and fish assemblages nationally. NAWQA integrated macroinvertebrate data (collected by NAWQA and the EPA Wadeable Stream Assessment[15]) to expand the scope of a model assessment of biological condition in streams in the western United States (Carlisle and Hawkins, 2008). These studies are the foundational material for a USGS Circular summarizing findings on aquatic communities across the United States prepared by the Ecological National Synthesis Project, planned for 2012.

---

[15] See http://water.epa.gov/type/rsl/monitoring/streamsurvey/index.cfm.

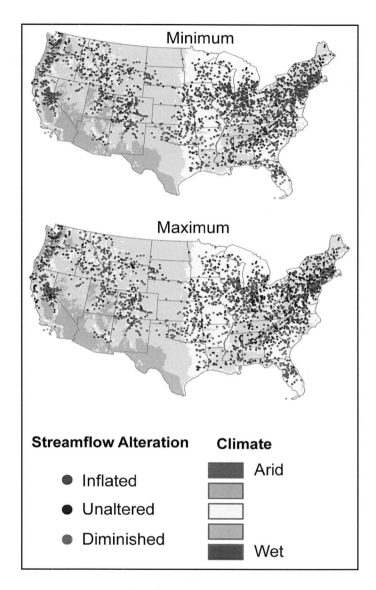

FIGURE 2-10 Alteration of minimum and maximum streamflow magnitudes at 2,888 sites monitored from 1980 to 2007. "Inflated" condition indicates that observed average magnitudes exceeded expected reference magnitudes. "Diminished" condition indicates that observed average magnitudes were less than expected reference magnitudes. SOURCE: Reprinted, with permission, from Carlisle et al., 2011. © 2011 by Ecological Society of America.

The SPARROW[16] model is NAWQA's most popular and visible regression model.[17] The SPARROW model is a watershed based model designed to predict patterns in water quality, concentration, and amount of constituents, across spatial extents ranging from entire regions of the United States to smaller watersheds. The model is perhaps best known for contributing to understanding of key parameters that affect hypoxia in the Gulf of Mexico by determining nutrient load to the Gulf and pinpointing which watersheds or which of the 31 state drainage basins are the greatest contributors. Specifically, the SPARROW effort highlighted that nine states[18] making up one-third of the Mississippi River drainage area contribute 75 percent of the nitrogen and phosphorus to the Gulf (Alexander et al., 2008). This study also filled gaps in the understanding on the sources of phosphorus in the Gulf; phosphorus associated with animal manure contributes almost as much phosphorus as cultivated crops (37 versus 43 percent) (Alexander et al., 2008).

Currently NAWQA is in the process of developing fine-scale, regional water-quality models in each MRB. Nutrients are the focus of these modeling efforts, except in the arid southwest, where dissolved solids are of greater importance. To do this, NAWQA is using local ancillary data and refining the SPARROW model to reflect the unique environmental conditions and smaller scale of each MRB. At this time, models have been developed for six of the eight MRBs. Regional models for the remaining basins, California and the Southwest, are planned for the future. The preliminary findings from this effort show the promise of future regional SPARROW modeling of water-quality conditions in the United States. The October 2011 issue of the *Journal of American Water Resources Association* provides a featured collection of articles on the regional SPARROW effort.[19]

NAWQA is exploring uncertainty in all the modeling efforts, i.e., associating uncertainty with all the estimates the program produces. For example, Robertson et al. (2009) examined approximately 800 watersheds in the Mississippi River basin and assigned a ranking that indicated whether nutrient yields from the basin were among the highest delivering of nutrients contributing to hypoxia in the northern Gulf of Mexico (Figure 2-11, top). This involved a robust statistical procedure applied to the results from a previous application of SPARROW to identify the top 150 watersheds. Once identified, scientists incorporated information on confidence intervals

---

[16] See http://water.usgs.gov/nawqa/sparrow/.

[17] Development of SPARROW was initiated by the Branch Systems Analysis working on new and emerging technical issues and techniques used within the former Water Resources Division. The branch was dissolved in the late 1990s because of funding shortfalls, and the individuals developing SPARROW joined NAWQA and continued their work.

[18] Illinois, Iowa, Indiana, Missouri, Arkansas, Kentucky, Tennessee, Ohio, and Mississippi.

[19] See http://onlinelibrary.wiley.com/doi/10.1111/jawr.2011.47.issue-5/issuetoc.

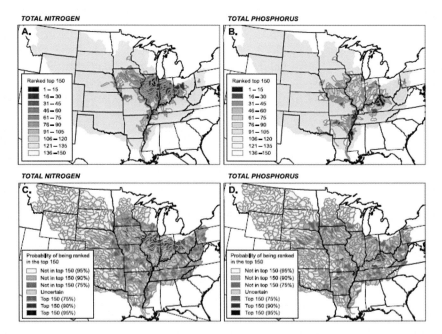

FIGURE 2-11 Map showing Total Nitrogen and Total Phosphorus (as delivered incremental yield) from the top 150 contributing watersheds (top). Map showing the certainty of placement within the top 150 contributing watersheds for Total Nitrogen and Total Phosphorus (bottom). SOURCE: Reprinted, with permission, from Robertson et al. 2009. © 2009 by John Wiley & Sons.

of these model predictions estimating the probability that these watersheds are among those that have the highest nutrient yields to the Gulf (Figure 2-11, bottom). This was a SPARROW spin-off project and was EPA driven. This information has important management implications for the Midwest and is being used by EPA to target non-point source pollution in those watersheds.

NAWQA is offering the use of monitoring and modeling tools to the user, an effort that will extend into Cycle 3. Although these efforts are still in their infancy, they represent a significant step forward for NAWQA and the user community. For example, the Watershed Regression for Pesticides models, referred to as WARP models, predict specific concentration statistics for a given pesticide in the United States. These models establish linkages between pesticides measured at NAWQA surface water sampling sites to variety of factors (pesticide use, soil characteristics, hydrology, and climate) that affect pesticides in streams. One of the first completed WARP models was for the pesticide atrazine (Larson and Gilliom, 2001), which was improved during Cycle 2 (Larson et al., 2004). Today, the atrazine WARP model and associated data are available for public use on the web.[20] The user can visit a website and see estimates of atrazine concentrations in an area or basin along with the error and uncertainty associated with that estimate. NAWQA scientists are planning to bring other pesticide data to the web in a similar fashion.

Another example of bringing modeling and monitoring activities to the user, the SPARROW Decision Support System provides online access to SPARROW models that can be used to predict long-term average water-quality conditions and source contributions by stream reach and catchment and to evaluate management source-reduction scenarios (Booth et al., 2011).[21] (For additional information see Box 4-1.) Also, USGS and EPA are working together to provide interested parties with a web service to assist in integrating large water-quality databases.[22] Users can go into the USGS website and retrieve data from the National Water Information System, which includes water-quality data from NAWQA, in a common format and go to the state EPA data (STORET) and retrieve data formatted in the same way.

CURRENT STATUS

Using the FY2011 appropriations for USGS as the metric, NAWQA's budget of $62.9 million was approximately one-third of the appropriation for water-related programs at USGS (the former Water Resources Discipline area). Although the allocation of the budget evolves with programmatic design, in FY2010 the majority of NAWQA's budget was used for program activities (for example, status and trends networks) versus program management or support of broader USGS efforts (Figure 2-12). The appropriations in actual or nominal dollars for NAWQA have been flat since the late 1990s or declining when adjusted for inflation (Figure 2-13). This has been consistent with the overall budget and staffing trends of water-related programs at USGS over the past 16 years, which are flat or declining (NRC, 2009).

NAWQA is visible to the public via the data and interpretive delivery systems the program strives to make publicly available, and the program has a record of scientific achievement since its inception (NRC, 1990, 2002, 2009, 2010, 2011a; USGS, 2010). NAWQA has produced approximately

---

[20] See http://infotrek.er.usgs.gov/warp/.
[21] See http://cida.usgs.gov/sparrow/.
[22] See http://qwwebservices.usgs.gov/.

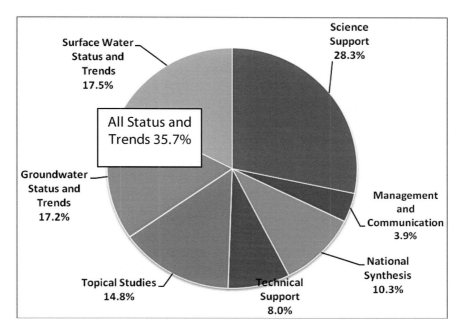

FIGURE 2-12 NAWQA funding by category in FY2009. Total appropriation for FY2010 was approximately $66.5 million. "Science Support" represents funds allocated for Bureau Science support (approximately 5 percent), the National Research Program (approximately 34 percent), and Water Mission Area Technical support (approximately 61 percent). The last supports the Office of Water Quality, the National Water Quality Laboratory including the Methods Research and Development Program, which develops new analytical methods, and the Branch of Quality Systems. "Technical Support" represents funds allocated to support the Hydrologic Systems Team, which provides modeling support to all components and includes the national SPARROW team and Data Synthesis Team, which provides data management support for NAWQA including the Data Warehouse and BioData database. "Management and Communication" represents funds allocated to support the NAWQA National Leadership Team and its support staff and NAWQA Communications staff. SOURCE: NAWQA National Leadership Team, personal communication, May 13, 2009.

1,900 reports during its 20-year history, a publication every 4.2 days on average, a value which, while not an indicator of quality, provides a sense of the quantity of work produced over the history of the program. (M. Larsen, personal communication, May 13, 2009). If released products are the metric (those already released and to be released), NAWQA has mined approximately one-third of the Cycle 1 data (NAWQA leadership, personal

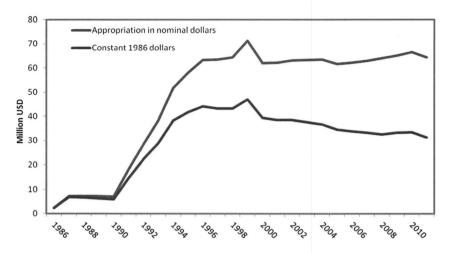

FIGURE 2-13 NAWQA appropriations history in nominal or non-inflation adjusted U.S. dollars (USD) and constant 1986 USD. Inflation was calculated using the Consumer Price Index inflation factor, and base year is an average across 1982-1984 and indexed at 100. SOURCE: FY appropriations from NAWQA leadership, personal communication, August 2011.

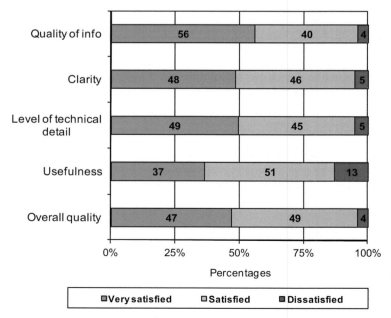

FIGURE 2-14 A Customer Satisfaction Survey, conducted in 2010, indicates user satisfaction with NAWQA information. SOURCE: USGS, personal communication.

communication, May 9, 2009), a value that, although not an indicator of quality, provides a sense of the quantity of work produced over the history of the program. A Customer Satisfaction Survey, conducted in 2010,[23] indicates that the majority of NAWQA users are either satisfied or very satisfied with NAWQA information (Figure 2-14).

The Statement of Task charges the committee to conduct an assessment of NAWQA's accomplishments. In response, the committee notes 10 representative accomplishments of NAWQA in Chapter 3 to answer the Statement of Task.

---

[23] The 2010 NAWQA Customer Satisfaction Survey, referenced several times in this report, was conducted by the USGS Office of Budget, Planning, and Integration. It was conducted in July and August of 2010 and consisted of a random sample of 500 persons from the NAWQA stakeholder database. The response rate to the survey was 37 percent.

3

Assessing Accomplishments
of the NAWQA Program

The backbone of the National Water-Quality Assessment (NAWQA) program has been the systematic collection and analysis of two decades' worth of chemical, biological, and physical water-quality data using consistent and scientifically sound methods at a national scale. The program provides the majority of the nation's information on the geographical occurrence of chemicals in the aquatic environment (streams, rivers, and groundwater). The first two decades of NAWQA's effort provide a record of accomplishment that is too extensive to present in detail. Therefore, this chapter identifies 10 representative accomplishments of the program (Box 3-1) and assesses their significance, thus "assessing the accomplishments of the NAWQA program" per the statement of task. The order in which they are presented does not represent an evaluation of their relative significance.

NATIONAL ASSESSMENT OF CHEMICALS IN
THE NATION'S SURFACE WATERS

Reports from individual study units in Cycle 1 established a baseline of water-quality in surface waters in distinct environmental settings with specific hydrogeology, climate, and anthropogenic factors. Data from all study units were combined to provide a national picture of NAWQA's water-quality findings (Hamilton et al., 2004), which revealed that although most water in the United States is fit for most uses, contamination from point and non-point sources affects every study unit, particularly agricultural and urban areas. Contamination is generally a complex mixture of nutrients,

---

BOX 3-1
Accomplishments of the NAWQA Program

National assessment of chemicals in the nation's surface water: NAWQA has provided a national picture of surface water quality.

National assessment of chemicals in the nation's groundwater: This picture extends to the quality of the nation's groundwater, giving the scientific and regulatory communities and the public an understanding of the nation's water quality. Specific to groundwater, NAWQA has demonstrated the utility of groundwater age determination in water-quality studies, especially mixing of old and young waters.

Incorporation of biological indicators of water quality into assessments: NAWQA has integrated measures of indicator organisms into water-quality monitoring and has examined relationships among biological, chemical, hydrological, and land-use parameters using uniform methods at a national scale.

National synthesis reports: These reports synthesize robust data sets using descriptive statistics to draw broad conclusions for the nation to help answer the question that led to the program's development—what is the state of the nation's water-quality?

Continuity and consistency in study methods and design: NAWQA uses standardized sampling regimes, network design, and analytical techniques to enable cross-site comparisons, as well as intensive site-specific and constituent-specific sampling to meet local and regional stakeholder needs, and national water-quality assessments.

---

pesticides, organics, and their breakdown products, which are often just as prevalent as the parent compounds. Contaminant occurrence is not limited to compounds currently in use: polychlorinated biphenyls, chlordane, dieldrin, and other organochlorine compounds that are now restricted still persist in streams and sediments. Spatiotemporal patterns in contamination correspond with the timing of chemical application, hydrologic events (e.g., snowmelt) and land management practices. Thus, NAWQA provided a picture of water quality nationwide, giving the scientific and regulatory communities and the public an idea of the nation's water quality.

NAWQA's continuing focus on pesticides built on the assessment of pesticides in the nation's surface waters and groundwaters from 1991 to 2001 (Gilliom et al., 2006). More recent analyses identify trends in pesticide and herbicide concentrations in streams and rivers in the Corn Belt from 1996 to 2006 (Sullivan et al., 2009). Regulatory and economic

Development and use of robust extrapolation and inference-based techniques: NAWQA has done an exemplary job of developing and applying robust extrapolation and inference-based models (e.g., SPARROW and the Watershed Regression for Pesticides or WARP models that are statistical, geospatial, and/or process-based and that support inferences from recent and historical data and projections of the outcome of proposed actions).

Information dissemination: NAWQA's communication activities have grown in scope and sophistication as the program has evolved. The program now uses multiple media and appealing graphics to communicate its information products and tools, and it has a wealth of publicly available water-quality data in its data warehouse.

NAWQA science informing policy and management decisions: The program has translated and interpreted its high-quality, nationally consistent data with sophisticated tools so that policy and decision makers can use the program's science to inform efficient decision-making.

Collaboration and cooperation: NAWQA continues to cooperate, coordinate, and collaborate within its own agency as well as with other federal, state, and local agencies in designing and carrying out its programs with a commitment to enhancing its usefulness by making its data and programs relevant to others with interests in water-quality.

Linkages and integration across media, disciplines, and multiple scales: NAWQA has been successful in multidisciplinary research at regional and national scales, collecting and interpreting geographic, hydrologic, biologic, geologic, and climatic data from a range of environmental media (e.g., groundwater, sediments, soils, surface waters, and biota) to help resolve water-quality questions.

changes caused major reductions in the application of some pesticides, with corresponding declines in surface water concentrations of those compounds (Gilliom et al., 2006). The NAWQA program's findings also highlight how the movement of nitrogen and pesticides from agricultural fields to streams, groundwater, and beyond is controlled by a complex yet identifiable interplay of hydrologic factors (irrigation, drainage, flow paths, precipitation), agricultural practices (compound applied, timing of application), and biological processes (photosynthesis, biological activity) (McCarthy, 2009).[1]

Water-quality improvements from reductions in pesticide use are not limited to agricultural areas. After a federally-mandated phase-out of the organophosphate insecticide diazinon in outdoor urban settings, the concentration in northeastern and Midwestern streams fell dramatically

---

[1] See http://in.water.usgs.gov/NAWQA_ACT/.

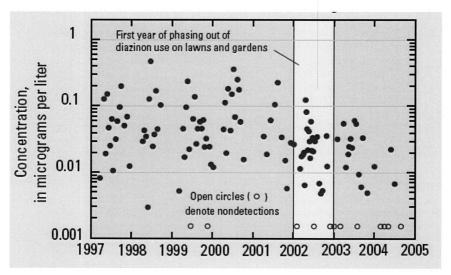

FIGURE 3-1 Diazinon concentration (μg/L) in Accotink Creek, Virginia (Potomac River Basin) from 1997 to 2005. As a result of the federally mandated phase-out of sales and use of the pesticide in 2001, concentration generally decreased after 2002. SOURCE: Gilliom et al., 2006.

(Figure 3-1), and the frequency of exceedance of the acute invertebrate water-quality benchmark (1 μg/L) in summer samples fell from 10 percent to less than 1 percent (Phillips et al., 2007).

To further advance the assessment of chemicals in the nation's surface waters, NAWQA scientists have used lake sediment cores to reconstruct water-quality histories. Accumulation rates for metals such as cadmium, chromium, copper, lead, mercury, nickel, and zinc have generally decreased since the 1970s, although accumulation rates in urban sediments can still be hundreds of times higher than rates in undeveloped watersheds (Mahler et al., 2006). Polyaromatic hydrocarbons (PAH) concentrations (associated with sediment) in cities where asphalt-based sealcoats are used are much lower than where coal-tar-based sealcoats are used (Van Metre et al., 2009). NAWQA also provides data gathering and sampling site assistance for work done by researchers in the U.S. Geological Survey (USGS) Toxics Program (Kolpin et al., 2002), which provided the first national-scale snapshot of the occurrence of contaminants of emerging concern.

Overall, NAWQA's surface water-quality monitoring efforts provide an invaluable data set of surface water-quality conditions across the nation. NAWQA uses these data to provide regional and national assessments of great value. Although publications detailing some Cycle 2 studies are still forthcoming, the program has already made significant steps toward being able to answer specific policy-relevant or national questions about surface water-quality.

A NATIONAL ASSESSMENT OF CHEMICALS IN U.S. GROUNDWATER

USGS has been successfully conducting groundwater studies for more than 100 years. Indeed, the "father of groundwater hydrology" was Oscar E. Meinzer, employed by USGS from 1910 to 1940 and who served as the third USGS Ground-Water Division Chief. USGS was the first governmental agency to systematically apply science to studying groundwater systems, and its regional assessments of groundwater resources remain a hallmark of how hydrogeologic synthesis is done, including the use of a broad range of USGS publicly available groundwater numerical and geochemical models, and other groundwater assessment tools.[2]

NAWQA's groundwater work builds on the USGS's strength in this field. NAWQA initially focused on how human activities affected groundwater quality in agricultural and urban areas, excluding considerations of surface water-groundwater interactions (NRC, 2002). However, the connection between groundwater and surface water makes it difficult to achieve understanding if the resources are treated independently, so NAWQA adopted a process-based approach during Cycle 2 to characterize and model surface water-groundwater interactions in all appropriate study units (NRC, 2002).

NAWQA has integrated groundwater elements into its studies, even those that do not specifically focus on aquifers. NAWQA's national synthesis reports on pesticides, volatile organic compounds (VOCs), and nutrients have all included important groundwater components. NAWQA's groundwater work is particularly important because little such work on groundwater quality has been done systematically at a large scale. For example as part of the national assessment of pesticides (Gilliom et al., 2006), NAWQA reported that 55 to 61 percent of shallow groundwater samples in urban and agricultural areas contained one or more pesticide compounds, compared with 29 to 33 percent of samples from undeveloped or mixed land use areas (McMahon et al., 2008). VOCs were detected in aquifers across the United States, although concentrations were below a detection threshold of 0.2 ppb in 80 percent of the wells (Zogorski et al., 2006).

The characterization of groundwater quality in regional aquifers builds upon a study conducted in the High Plains Aquifer (McMahon et al., 2007) designed to exploit existing data to improve understanding of regional groundwater quality and flow, particularly with respect to aquifer vulner-

---

[2] See http://water.usgs.gov/software.

ability to contamination (USGS, 2005). Modern agriculture and particularly irrigation practices have increased the concentration of nitrates and dissolved solids in shallow groundwater, especially in areas where the local hydrogeology is conducive to fast transport of chemical species. Water supply pumping schemes that mix deep and shallow groundwater often produce lower-quality water than those that pump from deep wells alone. Furthermore, changes in land use such as conversion of rangeland to irrigated crops can affect local shallow groundwater quality (Gurdak et al., 2009). This type of regional analysis connects groundwater characterization efforts on many levels (e.g., private wells, public wells, age dating, flowpath modeling) and enables informed management decisions (e.g., reducing the risk of groundwater contamination by supply well pumping schemes or decreasing transport of untreated runoff to susceptible topographic lows).

Through the Topical Study Contaminant Transport and Public Supply Wells, NAWQA scientists have demonstrated the utility of groundwater age distribution determination in water-quality studies, especially mixing of old and young waters (McMahon et al., 2008). NAWQA has further advanced the science so that particle-tracking numerical models can be used to generate age *distributions* of groundwater entering a public supply well, not simply groundwater *ages* (Ehberts at al., 2012). A public supply well with a high fraction of young water might indicate a susceptibility to contamination initiated by a land-use change, whereas a public supply well yielding very old groundwater might be less susceptible to that contamination source.

The efficacy of this approach was dramatically illustrated by McMahon et al. (2008), who studied public supply wells in four aquifers. The modeled water-quality response to measured and hypothetical land use changes was dependent upon the age distributions of groundwater captured by the public supply wells and upon the temporal and spatial variability in land use in the source areas contributing to the wells. The time scales for public supply wells water-quality changes could be on the order of years to centuries for land use changes that occur over days to decades. These findings have implications for policy- and decision-making in relation to source water protection strategies that rely on land use change to attain water-quality objectives.

The hyporheic zone is the interface between groundwater and surface water where an exchange of chemical species, water, and organisms occurs (Gibert et al., 1990; Vervier et al., 1992). The connection between groundwater and surface water dictates that even in surface water supply and environmental flow studies, some knowledge of groundwater quantity and quality is essential. NAWQA researchers have advanced knowledge of exchange processes in the hyporheic interface. Recent studies include an examination of denitrification in the hyporheic zone of low-gradient nutrient-rich streams (Puckett et al., 2008) and a demonstration of the usefulness of heat as an environmental tracer in surface water-groundwater quality studies, providing another tool for practitioners (Essaid et al., 2008). USGS pioneered hyporheic research, and now the importance of the hyporheic zone has been widely recognized and is currently being studied in research groups around the country (NRC, 2002).

INCORPORATION OF ECOLOGICAL ASSESSMENT INTO NAWQA

NAWQA scientists have integrated biological assessments into water-quality monitoring and have examined relationships among biological, chemical, hydrological, and land-use parameters using uniform methods at a national scale. More than 450 publications have resulted from NAWQA's ecological research,[3] although much of the work is still forthcoming. The ecological condition is being assessed with metrics derived from samples of algae, macroinvertebrates, and fishes, which is an important and unique aspect of the NAWQA data. It is rare that all three are assessed in monitoring programs, although NAWQA data reveal that the three types of organisms seldom exhibit similar degrees of alteration in response to different land uses (e.g., Cuffney and Falcone, 2009). This implies that assessments based on only one type of organism may misjudge the extent and severity of impairment. Additional findings are that hydrologic alteration and land use change are the major drivers of alterations in ecological condition.

Ecological work in Cycle 2 included topical studies and program efforts encompassing four research areas: effects of urbanization on stream ecosystems, effects of nutrient enrichment on stream ecosystems, mercury in stream ecosystems, and effects of hydrologic alteration. Some of this work is highlighted here.

NAWQA's Effects of Urbanization on Stream Ecosystems Topical Study studied how stream ecosystems respond physically, chemically, and biologically to urbanization, and how these responses vary in different geographic settings (Cuffney and Falcone, 2009; Giddings et al., 2009; McMahon, 2000; McMahon and Cuffney, 2000; Tate et al., 2005). NAWQA documented how regional patterns of development and regional differences in past and present land use (e.g., history of agriculture in the watershed) affect the response of biota to urbanization (Brown et al., 2009). Earlier researchers had suggested that the first signs of degradation appear when impervious surface cover reaches approximately 10 percent (Booth and Jackson, 1997; Schueler, 1994), but recent NAWQA studies found a continuous linear decline rather than a threshold (Brown et al., 2009; Cuffney

---

[3] See the NAWQA publication bibliography: http://water.usgs.gov/nawqa/bib/pubs. php?cat=2 (accessed October 2011).

et al., 2005). A predictive model has also been developed, allowing prediction of benthic invertebrate response to urbanization at basin or regional scales based on parameters that describe the environmental setting, including antecedent agricultural conditions (Kashuba et al., 2010).

The Topical Study Effects of Nutrient Enrichment on Stream Ecosystems examined the influence of natural and human-related factors on nutrient cycling in stream ecosystems in agricultural watersheds differing in crop types (row crop, orchard, vineyard, pasture), animals (beef and dairy cattle, poultry), irrigation practices (none, central pivot, furrows), tillage, and amount of fertilizer applied (Munn and Hamilton, 2003). These agricultural streams often have nutrient levels in excess of U.S. Environmental Protection Agency (EPA) nutrient guidelines and show a limited ability to remove excess nitrogen through algal productivity or denitrification, leading to elevated downstream transport of nitrogen (Duff et al., 2008; Frankforter et al., 2009). Recent NAWQA publications have begun to examine indicators and indices that could be used to relate nutrient conditions with biological conditions, land use, and habitat factors (Frankforter et al., 2009; Justus et al., 2010; Maret et al., 2010).

The past two decades have seen advances in scientific understanding of mercury occurrence and behavior in standing water bodies; in recent years, NAWQA researchers have made contributions to the state of knowledge through the Topical Study on Mercury in Stream Ecosystems. NAWQA has documented the occurrence and speciation of mercury in fish flesh, bed sediment, and stream water (Bauch et al., 2009; Scudder et al., 2009). NAWQA's analysis of recent and historical data for mercury in fish flesh (Chalmers et al., 2010) found that sites with decreasing trends in fish mercury outnumbered those with increasing trends by a factor of 6 between 1967 and 1987, demonstrating the effectiveness of the regulatory controls on mercury releases to the environment implemented during the 1970s. In a three-part article series NAWQA scientists described the chemistry and transport of mercury (Brigham et al., 2009) and contributed to understanding of the physical and biological factors that control the fate of mercury in stream ecosystems (Chasar et al., 2009; Marvin-DiPasquale et al., 2009). Drawing on multiple lines of evidence, the researchers concluded that the dominant factor controlling mercury concentrations in top predator fish is the amount of methylmercury available for uptake at the bottom of the food chain (Chasar et al., 2009).

The ecological effects of altered hydrology were studied using geospatial data to develop models predicting metrics of magnitude, frequency, duration, timing, and rate of change of streamflow (Carlisle et al., 2009). These models enable estimation of the natural flow regime, which is essential for estimating predisturbance conditions and for predicting natural flow characteristics at ungaged sites. A potentially significant quantitative tool for assessing ecological condition is the current effort to understand the relationship among land use, climate change, and streamflow alteration and to quantify relations between streamflow alteration and biological impairment (Carlisle et al., 2011).

Overall, the NAWQA ecology program has developed nationally consistent measures of the status of primary producers, macroinvertebrates, and fishes in rivers and streams. This has enabled a more complete and integrated assessment of the health of rivers and streams than would be possible with physical and chemical analyses alone. NAWQA's application of regression analysis and modeling of ecological data have facilitated identification of indicators and indices and may allow the development of predictive models. NAWQA's urban studies have contributed to the scientific community's efforts to advance integrative scientific understanding of urban streams (e.g., Wenger et al., 2009).

NATIONAL SYNTHESIS ASSESSMENTS AND REPORTS

NAWQA's National Synthesis Assessments and capstone reports use descriptive statistics to compare study unit data and other historical data (i.e., land use) to draw broad conclusions for the nation—a unique niche for NAWQA. National synthesis teams are able to write these reports because each NAWQA investigation adheres to a nationally consistent study design and employs uniform methods of data collection and analysis. NAWQA's ability to organize itself around these themes in contrast to a more traditional project-by-project approach represents a major organizational accomplishment. These reports help answer the original NAWQA question: what is the state of our nation's water quality? These reports identify water quality issues that occur only in isolated areas versus those that are pervasive, and they show the effects of human activities and natural factors on water quality in a range of environmental settings. Three national synthesis reports have been published (pesticides, VOCs, and nutrients), one is in progress (ecology), and the fate of the fifth (trace elements) is unclear.

The Pesticide National Synthesis Project[4] and corresponding national synthesis report, *Pesticides in the Nation's Streams and Ground Water, 1992-2001*, provides information about the occurrence of 75 pesticides and 8 pesticide degradates in agricultural, urban, undeveloped, and mixed land-use areas (Gilliom et al., 2006). Analytical methods "were designed to measure concentrations as low as economically and technically feasible," and results were assessed using human health, aquatic-life, and wildlife benchmarks. Pesticide concentrations in streams and groundwater were characterized by land use and geographic patterns in pesticide use as well

---

[4] See http://water.usgs.gov/nawqa/pnsp/.

as seasonal variations. Because of the 10-year sampling period, trends in concentration and aquatic life over time were detected and correlated to pesticide use.

The Volatile Organic Compounds National Synthesis Project[5] and corresponding national synthesis report, *The Quality of Our Nation's Waters—Volatile Organic Compounds in the Nation's Ground Water and Drinking-Water Supply Wells*, presents information about the concentrations of 55 VOCs in aquifers, considering factors such as geography, aquifer characteristics, VOC type, detection frequency, and well type (Zogorski et al., 2006). This information was used to examine associations between natural and anthropogenic factors and the 10 most frequently detected VOCs. Many of these VOCs are solvents and industrial chemicals that are of concern for aquatic and human health in drinking water sources. These associations should help federal, state, and local agencies design sampling programs to detect contamination.

The Nutrients National Synthesis Project[6] and corresponding national synthesis report, *Nutrients in the Nation's Streams and Groundwater*, describes nutrient occurrence, source, effects on humans and aquatic ecosystems, and trends in concentration between 1992 and 2004 (Dubrovsky et al., 2010). Median concentrations of total phosphorus and nitrogen in agricultural streams were six times greater than background levels. However, exceedence of the federal drinking water standard for nitrate as N (10 mg/L) is uncommon in streams used for drinking water and deep aquifers; this standard was exceeded in more than 20 percent of shallow[7] domestic wells in agricultural areas. Data for nitrogen and phosphorus show minimal changes in concentration in the majority of streams over the time frame studied, but more upward than downward trends occurred in those streams that did change in a statistically significant manner.

CONTINUITY AND CONSISTENCY IN STUDY METHODS AND DESIGN

In the late 1980s when discussions about a national water-quality assessment were gathering momentum, federal agencies could not answer the question of whether the 1972 Clean Water Act was producing the intended improvements in water quality nationwide (Knopman and Smith, 1993). A national-level water-quality assessment was not possible because of analytical inconsistencies and a multitude of sampling networks designed for other purposes and ultimately unsuitable for spatial or temporal compari-

---

[5] See http://water.usgs.gov/nawqa/vocs/.
[6] See http://water.usgs.gov/nawqa/nutrients/.
[7] Less than 100 feet below the water table.

sons. For example, USGS collected water quality data through the stream benchmark program in largely pristine small watersheds, in the National Stream Quality Accounting Network (NASQAN) program at the mouths of major river systems, and in its many cooperative study projects with states and local governments where sampling designs and constituents measured were largely problem-driven and particular to the place. EPA, states, and local governments collected water-quality data for monitoring and compliance purposes. Sampling at a given site was often started and stopped, depending on the project duration and funding, and hence few sites had sufficiently long records of consistent analysis to enable valid trend analysis at a national scale. Inconsistencies in data collection included differences in how a sample was taken from a stream, how the sample was handled after collection, and what analytical methods were used to measure chemical and biological constituents. A compelling original argument for NAWQA was USGS's ability to sustain a consistent, geographically diverse, and quality-assured data collection over decades, and follow through on a scientifically valid study design.

Since the program's infancy, NAWQA has standardized sampling regimes and network design to enable cross-site comparisons to meet local and regional stakeholder needs, but at the same time to enable a national water-quality assessment. NAWQA brought order to a wide range of practices and motivations in water-quality sampling and analysis. NAWQA uses USGS approved methods that have been developed and tested by USGS researchers and approved for use at a national scale. These methods are periodically published in the USGS *National Field Manual for the Collection of Water-Quality Data* (USGS, variously dated). NAWQA now provides a nationally consistent data collection and analysis of water-quality samples (Gilliom et al., 1995). In setting this example and working with other groups on consistent practices, NAWQA has also helped to improve the water-quality monitoring efforts of other entities. This is a significant and enduring accomplishment.

DEVELOPMENT AND USE OF ROBUST EXTRAPOLATION AND INFERENCE-BASED TECHNIQUES

NAWQA products are used to assess status and trends in water quality, to evaluate the effectiveness of regulatory programs, to inform policy analysis, and to support ecological risk assessment. For each of these applications, it is essential that data from a limited sampling of environmental attributes be put in geographical and climatic context with the uncertainty of inferences reported. NAWQA has developed and applied robust extrapolation and inference-based techniques that are statistical, geospatial, and/or process-based. These various models support inferences from recent and historical data and projections of the outcome of proposed actions. Using these techniques to define the quality of our nation's waters has added depth, both in space and in time, to the NAWQA assessment of U.S. water quality. Here two such models are highlighted, SPAtially Referenced Regressions on Watershed Attributes (SPARROW) and Watershed Regression for Pesticides (WARP).

The application of the SPARROW model (Smith et al., 1997) is an excellent example of USGS research and development leveraged by NAWQA. Although SPARROW was not developed under NAWQA, the program's extensive use and support for improvements has made the model increasingly valuable. SPARROW's capacity for quantitative evaluation of the origin, fate, and transport of contaminants in streams has pioneered a new way to investigate watersheds. SPARROW was designed as a national model to estimate long-term average values for water contaminants by relating in-stream water-quality and flow measurements with information about upstream sources and watershed characteristics. SPARROW assesses nutrient-source contributions, transport, and water-quality conditions at the national level, allowing estimation of nitrogen and phosphorus fluxes in unmonitored streams across the nation and enabling researchers to identify major nutrient sources and estimate nutrient fate in receiving bodies (Smith et al., 1997; USGS, 2009b). The model can be used to assess how large-scale changes in land use may affect future nutrient loading. NAWQA has refined the national model to study nitrogen delivery from the Mississippi River basin to the Gulf of Mexico (Alexander et al., 2000; Brezonik et al., 1999; USGS, 2009b).

NAWQA continues to transform SPARROW and explore this valuable tool. For example, the program is refining SPARROW to study various water-quality parameters in six of the eight Cycle 2 Major River Basins (MRBs). For future versions of SPARROW, NAWQA plans to incorporate updated geospatial and stream-monitoring data and to add temporal resolution that will facilitate analysis of decadal and seasonal change (USGS, 2009b). Also, NAWQA scientists are developing the SPARROW decision-support system, to bring the use of the model to the user through a USGS-supported web-based tool. (For more information, see Box 4-1 in Chapter 4.) SPARROW provides an important resource for evaluating and implementing management strategies; it integrates and benefits from data collected by collaborating agencies; and it is used by other organizations to help them meet water-quality objectives. Furthermore, it has the potential to contribute to all three of NAWQA's initiatives: status, trends, and understanding.

The WARP models are statistical/geographic information system models used to assess pesticide concentrations in unmonitored streams (Stone and Gilliom, 2009). To date WARP models have been used to probe agricultural applications of atrazine in streams, one of the most extensively used herbicides in the United States (Stone et al., 2008). Like SPARROW, these models, too, serve an important national purpose and may prove to be as useful as SPARROW in the future. For example, WARP models have recently been improved by developing region-specific models that include watershed characteristics that influence atrazine concentrations in the Corn Belt ("WARP-CB" models). The uncertainty for the regional WARP-CB models is lower than the national WARP models for the same sites (mentioned above and in Chapter 2), a promising development in terms of better prediction of atrazine in streams and for future WARP models of other pesticides (Stone and Gilliom, 2012).

INFORMATION DISSEMINATION

Effective communication of findings is critical to the success of a program like NAWQA and contributes to its perceived relevance and usefulness. This is because, as noted in NRC (2002), NAWQA is "first and foremost a provider of information to parties interested in water quality." Early in NAWQA's history, communication was promoted as a fourth unspoken NAWQA objective (apart from status, trends, and understanding). Since then, NAWQA communication activities have grown in scope and sophistication starting with the user-friendly, non-technical Delmarva Circular in 1991 (Hamilton and Shedlock, 1992). NAWQA has been a leader within USGS in developing new tools and approaches to communicating with its various audiences of federal agencies, local and state cooperators, public officials, and the general public.

NAWQA's communication activities have grown in scope and sophistication as the program has evolved so that these activities now represent one of the program's significant achievements despite the fact that a small percentage (1 to 2 percent) of the program's budget goes toward communication. In 2001 NAWQA released approximately 1,000 written publications. By January 2012 this number had grown to approximately 1,900, a publication every 4.2 days on average, a value that, while not an indicator of quality, provides a sense of the quantity of work produced over the history of the program. NAWQA is at an important junction in which key work for Cycle 2 is coming to completion and the program is launching a larger than normal amount of products as well as significant capstone products. NAWQA has 125 additional publications planned through 2012 as Cycle 2 draws to a close, pushing this total to more than 2,000 publications in the 20-year history of the program (Table 3-1).

Today, when NAWQA publishes a study, it produces a suite of publications and outreach activities according to a set communication plan designed to reach a variety of users. This communication plan uses a tiered

TABLE 3-1 Summary of NAWQA Publications by Type During the Pilot Phase, Cycle 1, and Cycle 2 through January 2012. More detailed information about publication types is included in Appendix C.

Scope and Primary Contents of Reports	Pilot (FY 1985–FY 1989)	Cycle 1 (FY 1990–FY 2000)	Cycle 2 (FY 2001–January 2012)	Additional publications to be completed in Cycle 2 (January 2012–the end of FY 2012)	Total
Circulars	3	35	33	19	90
Fact Sheets	0	175	75	4	254
Open File Reports	6	192	47	5	250
Water-Resources Information Reports	2	295	126	0	423
Conference Proceeding Papers	2	111	33	0	146
Journal Articles	2	188	301	76	567
Data Series Reports	0	1	26	4	31
Scientific Investigations Maps	0	0	4	0	4
Scientific Investigations Reports	0	0	138	15	153
Books, Chapters	0	12	8	0	20
Techniques and Methods Reports	0	0	3	0	3
Digital Media (audio, video, CoreCasts, yearbook)	0	2	14	1	17
Other (Professional Paper, Thesis, Water Supply Papers, Newsletters, Non UGSG Reports)[a]	1	25	18	1	44
Total	16	1036	826	125	2003

[a] Non-USGS reports indicates references produced outside of USGS that include either NAWQA data and/or are coauthored by NAWQA personnel, are about NAWQA, or are an interview with NAWQA personnel.

approach ranging from detailed scientific reports for technically trained audiences to one-page fact sheets and press releases for lay audiences. This includes informing the U.S. Congress; the program participated in approximately 25 congressional briefings throughout the history of the program (P. Hamilton, personal communication, May 13, 2009). Some of the work has been remarkably well cited in the scientific community, for example, *Effect of stream channel size on the delivery of nitrogen to the Gulf of Mexico* (Alexander et al., 2000) was cited 442 times as of August 20, 2012, according to Web of Science.

Perhaps the most notable strides in NAWQA's communication efforts during Cycle 2 were through the use of digital media and appealing graphics to communicate its information, products, and tools. NAWQA's home page is its primary web-based interface, which presents NAWQA publications, updates on recent findings, and links to project pages.[8] During Cycle 2, NAWQA improved the program's website by designing a more consistent look and feel to the individual web pages, improving access to information through national maps, creating web pages dedicated to individual topics, expanding related and embedded links through the site, and enhancing and expanding publication querying services. One notable example is the homepage of the Topical Study Contaminant Transport and Public Supply Wells.[9] Public use of the NAWQA website has increased since 2006, with most hits after release of reports (Figure 3-2); for example, there were approximately 60,000 requests after the release of SPARROW results listing the watersheds contributing most to nutrients in the Gulf of Mexico.

The USGS Office of Communication is developing a social media presence using a Facebook page, YouTube, and Twitter. This includes promoting NAWQA studies to the larger USGS audience, when appropriate. NAWQA and the USGS Office of Communication jointly develop video podcasts on various NAWQA studies as part of the USGS CoreCast series.

The NAWQA data warehouse[10] makes data widely available online with sufficient nodes for data approximating national coverage and, in some cases, with sufficient regional coverage to assess changes in water quality over time in major watersheds. It contains data on approximately 2,000 physical, chemical, and biological water-quality parameters (Bell and Williamson, 2006). Samples are from 7,300 stream and 9,800 wells as well as 3,000 bed sediment and tissue samples. Data include nutrient analyses for 66,000 samples, pesticide analyses for 44,000 samples, and VOC analyses for 12,000 samples. NAWQA biological information is

---

[8] See http://water.usgs.gov/nawqa.
[9] See http://oh.water.usgs.gov/tanc/NAWQATANC.htm.
[10] See http://water.usgs.gov/nawqa/data.

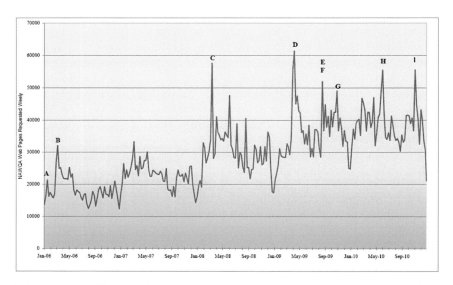

FIGURE 3-2 Public use of the NAWQA website since 2006 showing spikes with the release of the following studies and associated products: (A) Parking Lot Sealcoat: A Major Source of PAHs in Urban and Suburban Environments; (B) Pesticides in the Nation's Streams and Ground Water; (C) Water Availability—The Connection Between Water Use and Quality; (D) Ranking of SPARROW Model Nutrient Yields; (E) Mercury in Fish, Water, and Sediment; (F) Prediction of Atrazine Concentrations; (G) Agricultural Chemicals in Our Environment; (H) Effects of Urban Development on Stream Health; and (I) Altered Flows Leads to Ecological Degradation in Streams Across the United States. SOURCE: G. McMahon, personal communication, June 21, 2010.

available through the newly released BioData Retrieval System.[11] During Cycle 2, the data warehouse was improved with user-friendly mapping. The dissemination of NAWQA data via accessible databases enables scientists and regulatory agencies to place water-quality changes in geochemical and land-use contexts.

NAWQA SCIENCE INFORMING POLICY AND MANAGEMENT DECISIONS

NAWQA was created to support scientifically sound decisions for water-quality management, regulation and policy. NAWQA has translated and delivered its interpretation of program data to the policy- and decision-

---

[11] See http://infotrek.er.usgs.gov/nawqa_queries/jsp/biomaster.jsp.

makers who need it. Better science does not guarantee better policy, but NAWQA's data and scientific expertise inform efficient decision-making and thus have the potential to save resources. This is a significant program accomplishment. NAWQA tracks how its science and activities are used in decision-making and groups its contributions into 10 categories (Box 3-2).

Federal agencies including EPA, the National Oceanic and Atmospheric Administration, the Centers for Disease Control and Prevention, and the U.S. Department of Agriculture depend on NAWQA data for work on topics including nutrients, pesticides, stream protection and restoration, best management practices, fish consumption advisories, and even environmental factors related to nationwide cancer incidence. For example, the SPARROW model made substantive contributions to understanding of nitrogen and phosphorus sources and transport in the Mississippi River basin (Alexander et al., 2008). The study has major implications for "dead zone" hypoxia in the Gulf, and it will continue to help scientists and policy makers develop cost-effective nutrient management and reduction strategies in more than 800 watersheds within the largest drainage basin in the nation (USGS, 2010). Indeed, the federal interagency Mississippi River/Gulf of Mexico Watershed Nutrient Task Force is using this and other informa-

---

BOX 3-2
The NAWQA Program's Science and Activities
That Support Policy and Management

- "Assessing sources and transport of contaminants in agricultural and urban areas;
- Assessing vulnerability to help prioritize geographic areas, basins, and aquifers for management and protection;
- Understanding trends and whether conditions are better or worse over time;
- Assessing source-water quality used for drinking;
- Assessing and sustaining aquatic ecosystem health;
- Linking tributaries to receiving waters;
- Support for the development of regulations, standards, guidelines, and criteria for contaminants;
- Contributions to state assessments of beneficial uses and impaired waters (Total Maximum Daily Loads or TMDL), strategies for source water protection and management, pesticides and nutrient management plans, and fish-consumption advisories;
- Improved strategies and protocols for monitoring, sampling, and analysis;
- Communication of findings for policy and management."

SOURCE: USGS, 2010.

tion to make recommendations for action in the basin. More than 10 states and tribes use NAWQA data to meet EPA requirements, especially related to Total Maximum Daily Loads (USGS, 2010). The use of SPARROW also extends to understanding sediment loading in the Chesapeake Bay (Brakebill et al., 2010) and the sources of salinity in the southwest (Anning et al., 2007).

Decision-making, regulatory, and advisory bodies from local councils to state legislatures in more than 30 states also use NAWQA science to the benefit of public health and water resource management (USGS, 2010). NAWQA's work has enabled improvements in areas such as source water protection, quality assurance, quality control, sampling design, sampling methods, analytical protocols, and interpretation frameworks for the water resources issues that states and local governments confront. States save resources by using NAWQA data for these purposes. Washington and New Jersey have both used NAWQA data to obtain compliance monitoring waivers from the EPA for low vulnerability water supply wells under the Safe Drinking Water Act. Organizations like the Wind River Environmental Quality Commission of the Shoshone and Arapahoe Tribes in Wyoming use NAWQA's data to meet federal reporting requirements.

COOPERATION, COORDINATION, AND COLLABORATION

NAWQA has a history of cooperating and collaborating within its own agency, the Department of the Interior, and with other federal, state, and local agencies in designing and carrying out its programs. Those efforts to establish cooperative relationships have been recognized in past reviews (NRC, 2002, 2009). The following assessment from NRC (2002) remains true today:

> NAWQA has become a model of an effective, collaborative federal program—an attribute policy makers always stress, but seldom achieve. NAWQA has successfully integrated its program both within and outside of the USGS, establishing some exemplary relations with EPA and state governments.

NAWQA has continued to improve its efforts in this area during Cycle 2. NAWQA sites are coordinated with USGS's National Stream Accounting Network (NASQAN), thus strengthening the program's surface water network from within the agency. One particularly noteworthy product of external collaboration resulted from combining data from EPA's Wadeable Streams Assessment (WSA) and NAWQA to develop predictive models that provide taxon-specific measures of probability of capture, which were used to assess the biological condition of streams in several land use categories (Carlisle and Hawkins, 2008). The addition of NAWQA reference sites to the WSA model increased the range of environmental conditions to which the model could be applied.

Collaboration with the National Research Program and the Toxic Substances Hydrology Program have provided NAWQA scientists with new analytical methods, assistance in model development, and access to the latest insights from basic research. The committee heard from representatives of federal agencies (e.g., several program offices in the EPA, the U.S. Fish and Wildlife Service), states, and non-profits (e.g., the H. John Heinz III Center for Science, Economics and the Environment) that testified to the productive and collaborative relationships they have developed with NAWQA. Input from collaborators was essential to the development of the Cycle 3 Science Plan. These and other examples in Chapter 5 illustrate NAWQA's ability to collaborate with other programs and its commitment to enhancing its usefulness by making its data and programs relevant to others with interests in water quality.

LINKAGES AND INTEGRATION ACROSS MEDIA, DISCIPLINES, AND MULTIPLE SCALES

NAWQA has been successful in multidisciplinary research at the regional and national scales, integrating geographic, hydrologic, biologic, geologic, and climatic data, to resolve water-quality questions. NAWQA has collected and interpreted data from a range of environmental media including groundwater, sediments, soils, surface waters, and biota and focused attention on linkages between groundwater and surface water. NAWQA investigations consistently recognize the interrelatedness of processes occurring in aquatic and terrestrial environments that impact water quality. For example, NAWQA's work on mercury spans media including water column and suspended particulate matter (Brigham et al., 2009), sediment pore water (Pasquale et al., 2009), and fish and invertebrates (Chasar et al., 2009). These studies permit a holistic assessment of the complex dynamics and impact of mercury at the ecosystem scale.

NAWQA has successfully linked the disciplines of surface water and groundwater hydrology, chemistry, and biology. It is only through this multidisciplinary approach that the complexities of contaminants that cycle (e.g., metals and nitrate) and their impact on biota can be determined or that the relation between hydrology and contaminant transport can be quantified (Tesoriero et al., 2007). Because NAWQA designs, implements, and interprets study data with teams consisting of hydrologists, chemists, and biologists, the resulting reports offer cohesive and high-impact information on the complex interactions between chemicals and the physical and biological media through which they pass and interact.

NAWQA is uniquely positioned to collect and interpret data from scales ranging from single rivers and watersheds (Duff et al., 2008) to larger basins and aquifer systems, and finally to the entire nation (Lapham et al., 2005; USGS, 2008). Most NAWQA studies are conducted in systems that cross political boundaries (e.g., Alexander et al., 2008) and over time scales that range from short term (days to months) to years (Van Metre and Mahler, 2005) and decades (Mahler et al., 2006). NAWQA continuously and consistently collects and interprets data over time scales that are relevant to hydrogeologic processes and the impact of human activities on them. Studies mentioned in earlier sections have benefited from the enhanced spatial (e.g., urban stream studies) and temporal (e.g., principal aquifer studies) perspective. NAWQA is uniquely positioned to carry out complex, interdisciplinary work at scales that are not possible to achieve by individual academic or government scientists.

CONCLUSION

NAWQA has achieved a national water-quality assessment. This judgment is based on the committee's review of NAWQA achievements (Chapter 3), stakeholder assessments of the program heard in testimony, information contained in the NAWQA Science to Policy Management document (USGS, 2010), and the results of two NAWQA Customer Satisfaction Surveys (mentioned in Chapters 4 and 5). **The committee concludes that in Cycles 1 and 2, NAWQA provided a successful national assessment of U.S. water quality, in accordance with the mission of a national water-quality assessment program.**

4

The Way Forward for the Third Decade of National Water-Quality Assessment

Reasons to support the National Water-Quality Assessment (NAWQA) program in the third decade echo those that originally motivated the creation of the program. Indeed, the needs articulated by the National Research Council (NRC) in 1987 (NRC, 1987), needs that NAWQA was designed to meet and has met, remain ongoing and unchanged:

- characterize water quality over time,
- develop tools to evaluate why water quality has changed,
- provide water-quality data comprehensively to the nation in an accessible form,
- understand aquatic ecosystems, and, ultimately,
- forecast water-quality changes in the future.

More than 20 years after the 1987 NRC report was written, an additional reason to support NAWQA is its record of success and impact. Furthermore, as water-related issues become more complex because of changing climate, land use, and demographics the continued need for a national water-quality assessment becomes even clearer. Water-quality impairments will continue to be a complex issue, and resolving water policy debates will require more water science, not less (NRC, 2009).

The beginning of Cycle 3 is when the program can begin to achieve a new level of understanding and analysis capability even as it continues to document the status and trends of the nation's water-quality. NAWQA has evolved from a program emphasizing water quality data collection and trend assessment to one having the potential to forecast contaminant occurrence and aquatic degradation trends under multiple scenarios at nationally significant scales. In other words, NAWQA is poised, both within the U.S. Geological Survey (USGS) and the federal government, to understand the interplay between the complex factors that affect water quality through the continued requisite sampling of the nation's waters (NRC, 2011a). The program's scientific investments are maturing, enabling NAWQA to move beyond water-quality monitoring toward understanding the dynamics of water-quality changes and using that understanding to forecast likely future conditions under different scenarios of change. These are advances that the nation needs and the committee strongly supports (NRC, 2011a). **The need for a national water-quality assessment is as important, *if not more so today*, as it was when NAWQA was first established.**

A successful national water-quality assessment in Cycle 3 would be a national-scale water-quality surveillance program that evaluates and forecasts how changing land use conditions and climate variability may affect water quality in different settings, and that serves as a tool for water policy- and decision-makers as they evaluate policy options impacting the nation's water resources. Many efforts exist to assess water quality in the United States at universities and other federal and state programs at the local and regional levels. As the nation's water-quality regulator, the U.S. Environmental Protection Agency (EPA) has a particularly critical role. However, NAWQA is unique in its focus on water-quality assessment at the national scale and its inclusion of a large number of water-quality parameters. This corresponds with the committee's sense of the unique niche of a national program, a program that takes on work that states cannot do alone or work that crosses jurisdictional boundaries. For example, NAWQA would take on regional studies that can be extrapolated to other areas of the country, or studies that answer regional water-quality questions that are extremely important to the nation. Program efforts would cross state lines, such as water quality assessments of the Mississippi River.

Yet it is unrealistic to consider a way forward while ignoring fiscal realities and the difficult programmatic decisions that NAWQA will face. The committee sees many challenges ahead for NAWQA in Cycle 3, challenges that are related to the Statement of Task:

- How does NAWQA remain a national program in the face of resource decline?
- How should NAWQA balance new status activities against the need to maintain long-term trend networks and understanding studies?
- How can NAWQA use ancillary data and maintain a high level of quality?
- How can NAWQA maintain focus amidst numerous and competing stakeholder demands?

Chapter 3 should serve as a reminder to the program of a deep history of success to draw upon as it faces the challenges listed above. This chapter is framed in terms of priorities and trade-offs in order to be the most useful to the program and USGS.

THE FIRST PRIORITY: BASIC SAMPLING

NAWQA has produced a rich national database of chemical, physical, and biological water-quality information that covers a diverse range of water resources through a robust monitoring design. These data are essential for assessing the status and trends of the nation's water quality and are used by a large and varied number of stakeholders from other federal agencies to citizens. These data are used to develop, calibrate, and validate models that allow USGS and others to forecast future conditions under a variety of scenarios and extrapolate specific data points in order to a complete a "picture" of a given condition. NAWQA's basic sampling networks are critical.

Why does the nation continue to need long-term monitoring? Monitoring over many years to decades is critical to assess whether the quality of the nation's waters is improving or degrading, because of lag times in environmental responses and year-to-year variability. Monitoring is also essential to assess whether management strategies are working to improve water systems in, for example, the Chesapeake Bay (NRC, 2011b) or California's San Francisco Bay Delta Estuary (NRC, 2011c). Long-term, continuous collection of water-quality data serves an even broader-scale purpose by identifying changes in water quality caused by changes in the landscape condition, contaminant sources, and variations in climate. Calibrating water-quality models requires measures of both quantity and quality, along with a fundamental understanding of chemical and biological processing. Models that are produced to make predictions can only be validated through monitoring. Despite these and other reasons to support the need for long-term monitoring, observational networks to measure various water-quality characteristics in the United States have been on the decline for a number of years (Entekhabi et al., 1999; NRC, 1991, 2002, 2004b). It is important that NAWQA continue to help determine if policy changes related to water quality have been effective, particularly with respect to delivery of excess nutrients and contaminants to water supplies and important ecosystems.

The continuity of national water-quality measurements in time and space is fundamental to meeting the goal of national water-quality assessment and is something that no other entity in the United States has the capacity or charge to do. **First and foremost, NAWQA's primary focus should be on continuing the monitoring needed to support the national status and**

trends assessments of the nation's water-quality. Budgetary constraints and the need to fulfill the primary mission of the program make this focus even more critical. Once lost, such a perspective can be very difficult to reestablish resulting in a "break" in the long-term status and trends data set that NAWQA has established. Also, if basic monitoring data collection is reduced too far NAWQA could fall below the tipping point where it can be considered a national program in scope. This has been discussed in previous NRC reviews of the NAWQA program, noting that NAWQA could be nearing this tipping point where it is no longer a national program (NRC, 2002). Thus, the committee supports efforts in Cycle 3 that not only reach beyond the focus of basic monitoring, discussed below, but also recognizes that these other goals can only be accomplished if the basic data collection continues.

THE ROLE AND NEED FOR MODELING

A tipping point for NAWQA is a point where, once crossed, the program as currently organized, scaled, and operated can no longer provide a national assessment of water quality. Restoration of resources will not reverse this inability to achieve the program's core mission, once the tipping point is crossed. Scaling the program up to what it once was would be inhibited by the break in the long-term monitoring record and the erosion of programmatic infrastructure. However, there may be other scales, modes or organization, and scientific effort that would still allow water-quality monitoring to be achieved. Yet this water-quality monitoring would lack a key feature of the program—national scale—or the ability to say something meaningful about the nation's water quality as a whole.

The committee cannot quantify an exact tipping point for NAWQA. Metrics for identifying the point at which the tipping point is crossed, perhaps built into the network design, would be required. However, the committee can reflect on how to assess proximity to the tipping point through the critical question, how much could uncertainty increase in NAWQA outputs before relevant national conclusions could no longer be drawn, and the program suffered irreparable harm? Similarly, does NAWQA have adequate water-quality monitoring data to support its water-quality models?

Measurements can only provide a snapshot of condition for the time they are taken, and they cannot be used by themselves to forecast future conditions or understand water quality in unsampled areas. Models are tools that can be used for forecasting, as well as to construct scenarios for assessing the impacts of climate and land use change, and likely consequences of different policy options. **A focus of NAWQA efforts in Cycle 3, second only to basic monitoring activities, should be support of NAWQA modeling initiatives.**

NAWQA water-quality models initially are calibrated, matched—to data collected in the present—and usually used to "forecast" or to determine what trends would occur under different scenarios of demographic, land use, and climate change in order to address national issues and extrapolate to a national picture. This includes but is not limited to new initiatives involving the Watershed Regression for Pesticides (WARP) and SPAtially Referenced Regressions on Watershed Attributes (SPARROW) models (Box 4-1). These same models can be used, if desired, to "backcast," or to start with defining what water quality is desired in the future, and then identify what actions would control, say, nutrient loading to achieve that end result. These modeling and decision-support tools need to be accessible to researchers, water managers, and policy makers.

Land use and climate change call into question the efficacy of using historical data to assess hydrologic and ecologic conditions because both introduce nonstationarity into the hydroclimatologic record (Milly et al., 2008). Reconciling changing factors and trends in key observed variables is an important challenge for the detection and attribution of change. Reconciling nonstationarity will be challenging for models like SPARROW and WARP—indeed, for any model using historical data.

ASSESSMENT OF THE CYCLE 3 SCIENCE PLAN

The Science Plan provides a forward-thinking vision (Box 4-2) for the next decade of assessing the nation's aquatic resources. The Science Plan reflects many recommendations and suggestions from this committee's two letter reports. It outlines a well-connected conceptual model for the program in Cycle 3 linking status and trends to understanding sources of stressors and effects and then ultimately linking this to modeling efforts. The Science Plan is organized into four goals for the program, which constitute the logical maturation of the program and are wise choices for leveraging the previous two decades of monitoring.

The committee lacks the specifics to probe in great detail the technical soundness of specific methodologies and technologies to be used in Cycle 3; the Science Plan is a high-level planning document, and many details were not included. However, to respond to the Statement of Task,

> Review strategic science and implementation plans for Cycle 3 for technical soundness and ability to meet stated objectives.

an assessment of the technical soundness of the Science Plan and its ability to meet stated objectives follows.

The overall scope of the Science Plan is broad. Opportunities exist for NAWQA to gain efficiencies by reaching out to a broader technical com-

---

BOX 4-1
An Evolving SPARROW in Cycle 3

Previous versions of the SPARROW model were calibrated and used at a national scale to assess nutrient conditions in surface waters across the United States. Currently, NAWQA is expanding the use of the SPARROW model by calibration to encompass new scales and contaminants and by making all SPARROW models available to the public. For example, to obtain a more accurate assessment of water-quality conditions, NAWQA is currently calibrating the SPARROW model to six of the eight Major River Basins of the conterminous United States.

NAWQA is exploring the types of contaminants that can be modeled by SPARROW. The program is developing a national-scale organic carbon model that will simulate the national carbon balance. This naturally leads the SPARROW modeling effort to dissolved oxygen in surface waters and to a national-scale temperature model, the two of which are conceptually linked because dissolved oxygen responds quickly to temperature. Program scientists are also contemplating a national-scale dissolved solids and salinity model, which will have numerous practical applications, for example, in tracking the presence and impact of deicing road salt.

Finally, the SPARROW Decision Support System (DSS) is a new, USGS-maintained repository for SPARROW models that are made available for public use. This tool makes SPARROW available to the public via a USGS web-based system.[1] The DSS allows users to choose a desired model, craft water-quality scenarios, manipulate the models locally, and share and upload the information at a later date. The system has a mapping interface that can be manipulated to show a variety of results. Datasets that calibrate the models will be available as well.

When considering these developments together—the expansion of contaminants, scale, and bringing SPARROW to the public through the DSS—it is clear that NAWQA personnel envision an enhanced SPARROW for the future. This is a vision the committee supports.

---

[1] See http://cida.usgs.gov/sparrow/#modelid=53.

---

munity for innovation, monitoring, and analysis. Thus, with respect to Statement of Task 1b,

> Are there issues not currently being substantially addressed by NAWQA that should be considered for addition to the scope of NAWQA?

The committee recommends that no other issue(s) should be considered for addition to NAWQA in Cycle 3. NAWQA has identified the major water-quality issues facing the nation in the Science Plan that fall within its purview.

BOX 4-2
The Guiding Vision for Cycle 3

"Science-based strategies can protect and improve water quality for people and ecosystems even as population and threats to water quality continue to grow, demand for water increases, and climate changes."

SOURCE: Design of Cycle 3 of the National Water-Quality Assessment Program, 2013-2023: Part 2: Science Plan for Improved Water-Quality Information and Management

---

Effectiveness of Presentation

The opening chapter of the Science Plan provides a compelling description of NAWQA's vision for Cycle 3. The chapter continues to successfully articulate how NAWQA is uniquely positioned to address some of the nation's most pressing water-quality issues, including an assessment of the nation's water quality and the stressors that place water quality at risk of decline. The connections among the four goals of Cycle 3 are clearly articulated, as are the benefits of the Cycle 3 plan to the nation. The chapter explains why Cycle 3 is needed now and how partnerships are needed to address the nation's need for clean water and to address problems due to shifts in population, changes in land use, and climate change.

However, the subsequent chapters of the Science Plan that expand on the main themes presented in Chapter 1 could be more clear and succinct. As a result, the committee's positive impression of the Science Plan comes more from the first chapter of the Science Plan and from presentations given by the NAWQA leadership team during the committee's deliberations rather than from the more detailed chapters in the Science Plan. More specifically, the presentation of the goals is unbalanced; Goal 1 is very long and provides significant detail on each subobjective (30 pages), while Goals 2-4 are described by far less text and appear less well-developed. In addition, redundancies exist among the chapters and ultimately detract from the message. The description of each goal lacks the requisite preface needed to identify the data gaps addressed by the activities described for each goal (e.g., model inputs). Subobjectives are not prioritized in their order of appearance (Goal 1).

Although the committee is confident in the overall Science Plan and the direction for Cycle 3, in places the presentation and development of the written document does not instill the same level of confidence. This point is made not to be prescriptive about specific revisions to the Science Plan, but to encourage NAWQA to continue to be mindful of its presentation of the Science Plan and even the forthcoming Implementation Plan. These are planning documents of a more internal nature; yet, a correlation to program impact does exist in any public document that NAWQA produces. This is particularly true in documents guiding the vision for the future.

Linking Groundwater and Surface Water

NAWQA plans to, in part, assess groundwater quality as a source of drinking water in Cycle 3. Although understanding the contamination of the nation's source water for drinking water supply is important, this coverage of and primary focus on a single use seem inadequate to meeting the stated NAWQA mission (see also, NRC, 2010, Appendix B). Understanding groundwater flows and articulation of the interconnectedness of groundwater and surface water is an important theme. For years, USGS and NAWQA have been educating the scientific community and the public about this relationship and have been conducting seminal research to establish and explain these relationships. The committee concludes that NAWQA should be mindful of this role in Cycle 3.

THE SCIENCE PLAN GOALS AND OBJECTIVES: AN EVALUATION OF TRADE-OFFS

The Science Plan is structured around four goals, each of which:

> relate to the underlying program principles of status, trends, and understanding. These goals are: 1) Assess the current quality of the Nation's freshwater resources and how water quality is changing over time; 2) Evaluate how human activities and natural factors, such as land use and climate change, are affecting water quality over time; 3) Determine the relative effects, mechanisms of activity, and management implications of multiple stressors in aquatic ecosystems; and 4) Predict the effects of human activities, climate change, and measurement strategies on future water quality and ecosystem condition (Design of Cycle 3 of the National Water Quality Assessment Program, 2013-2023: Part 2: Science Plan for Improved Water-Quality Information and Management).

The four goals in the Science Plan are consistent with the guiding vision, and they contribute to meeting the vision in a synergistic, interconnected, and balanced manner (although not communicated equally well, as noted above). The goals are used to guide development of activities that address priority "stressors" and their impact on water quality (Figure 4-1).

Then, the Science Plan lists 20 objectives under the auspices of the four main goals that outline the scientific work planned to achieve each goal. The several specific objectives that are described under each of the four

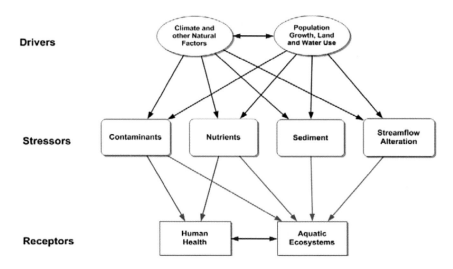

FIGURE 4-1 The four major water-quality stressors (contaminants, nutrients, sediment, and streamflow alternation), all of which are impacted by large-scale drivers (climate and other natural factors and population growth, land and water use) that guide the Cycle 3 Science Plan goals and program activities. SOURCE: Design of Cycle 3 of the National Water Quality Assessment Program, 2013-2023: Part 2: Science Plan for Improved Water-Quality Information and Management.

goals in the Science Plan are not necessarily equal in their contribution to meeting the central or core principles of the Science Plan, or to meeting the overall program mission. Not only do these various objectives differ in their potential impact and in their contributions to the programmatic goals, but also they differ in the effort and resources they will require, the clarity of how they are presented, how well they are justified, and the consequences of pursuing them with higher or lower priority.

The committee provides a discussion of the goals and objectives in the Science Plan to help inform NAWQA as it moves forward into Cycle 3 and adapts to changes in the future, speaking to Statement of Task 1. The committee is not charged with assessing the budgetary dimensions of NAWQA's goals and objectives as part of this review of the program, nor is it qualified to do so. Yet to be sensitive to the impact of available funding on programming, this guidance is provided as a discussion of trade-offs and consequences should the funding for implementing the Science Plan not be provided in full, with the overarching purpose of ensuring that NAWQA remains a national program.

Considerations Used in the Evaluation of Trade-offs for Cycle 3

The Cycle 3 Science Plan offers a comprehensive assessment of the nation's needs for understanding status and trends in surface and groundwater quality and developing a portfolio of multi-scale models to forecast changes in water quality in response to changes in demography, land use, and climate. The Science Plan articulates an ambitious agenda of 4 goals and 20 objectives that drive data collection, model development, and products for public dissemination.

It is critical to keep in mind that Cycle 3 should build on the existing two decades of data, experience, and products. The committee believes the Science Plan does that well. It is also important to keep the perspective that the Science Plan for the coming decade is important not only to NAWQA, or to USGS or the Department of the Interior, but also to the nation. The federal government will not be able to answer the question "Is the nation's water quality getting better or worse?" without NAWQA. In an ideal world, the Cycle 3 Science Plan would be implemented in full. All 20 objectives have scientific merit. However, given the current federal fiscal climate and the scale of the Science Plan, full-scale implementation of the Science Plan is unlikely.[1]

The committee carefully considered the Science Plan objectives in light of NAWQA's mission, capabilities, and resources and considered whether objectives were critical to the program mission and associated trade-offs. The committee developed criteria for determining which objectives are "essential" to NAWQA as a national program. An objective is essential if it contributes to one or more of the following:

- monitoring status and trends of surface and groundwater quality and relevant aquatic ecosystem indicators on a national scale;
- providing modeling capabilities to understand the effects of multiple water quality stressors on humans and ecosystems, and the impacts of climate change, land use practices, and demographic changes;
- assessing regional-scale effects of climate change, changing land use practices, and demographic changes;
- forecasting consequences of future scenarios with regional (multistate) and national implications.

Implicit in the consideration of "essential" modeling objectives is that NAWQA would embed rigorous model validation processes preceding full-scale deployment of models.

---

[1] This supposition is derived from conversations with NAWQA leadership and a set of fiscal scenarios crafted in the Science Framework. These scenarios estimate low, moderate, and high funding levels (compared to fiscal year 2009 levels) and the correlation to what activities the program could pursue in Cycle 3.

In a second category, the committee identified objectives that can provide important benefits to the nation and would have consequences if they were not accomplished, but are not essential to NAWQA's achievement of its core mission as a national water-quality program (i.e., "important but not essential"). In some cases, these objectives are being addressed by other entities. For these objectives, the committee believes that NAWQA should play a contributing role, or work closely in partnership with other organizations with complementary capabilities. The discussion identifies partners; Chapter 5 explores this further. In other cases, despite scientific merit, the committee concluded that the magnitude of resources necessary to achieve the stated objective would detract from other, program-critical, goals.

Finally, in a third category, the committee identified one objective for which the Science Plan does not provide sufficient justification for its value to the nation and its place within NAWQA (i.e., needs "further justification"). Consequently, the objective in this third category is of least importance to the program. This determination, along with the categorization of the other objectives, does not imply that this objective is without scientific merit; quite the opposite. The following discussion is a practical exercise in an attempt to assist NAWQA leadership in making difficult decisions regarding future priorities.

Below, the committee discusses the placement of the Science Plan objectives in these three categories, beginning with the first. Many objectives in the Science Plan overlap both conceptually and in how NAWQA will achieve the goal in the Cycle 3 program design. Thus, the discussion is framed not only around objectives that are "essential," "not essential," and also need "further justification" but why the scientific thrusts embedded within each are particularly critical.

Objectives That Are *Essential* to Cycle 3

The committee advises that these objectives are crucial to NAWQA's mission and to remaining a national program.

Goal 1 Status and Trends: Objectives "a" (surface water), "d" (groundwater), "e" (stream ecosystems), "f" (contamination of receiving waters), and "g" (biological condition)

The essence of this goal is the very reason the program was established: the need to develop long-term, nationally consistent information on the quality of the nation's streams and groundwater (Box 4-3). The data and analyses associated with the Goal 1 objectives continue the original NAWQA objectives of assessing the status and trends of the nation's water quality and the factors that affect water quality and aquatic ecosystems. In

---

BOX 4-3
NAWQA Science Plan, Goal 1 and Objectives

Goal 1: Assess the current quality of the Nation's freshwater resources and how water quality is changing over time.

Essential Objectives:
1a. Determine the distributions and trends of contaminants in current and future sources of drinking water from streams, rivers, lakes, and reservoirs.

1d. Determine the distributions and trends of contaminants of concern in aquifers needed for domestic and public supplies of drinking water.

1e. Determine the distributions and trends for contaminants, nutrients, sediment and streamflow alteration that may degrade stream ecosystems.

1f. Determine contaminant, nutrient, and sediment loads to coastal estuaries and other receiving waters.

1g. Determine trends in biological condition in relation to trends and changes in contaminants, nutrients, sediment, and streamflow alteration.

Important but Not Essential Objectives:
1c. Determine the distributions and trends in microbial contaminants in streams and rivers used for recreation.

Objectives That Need Further Justification:
1b. Determine mercury trends in fish tissue.

SOURCE: Design of Cycle 3 of the National Water Quality Assessment Program, 2013-2023: Part 2: Science Plan for Improved Water-Quality Information and Management; January 28 2011.

---

addition, the data collected for this goal are needed to meet the objectives and other Cycle 3 goals and will contribute much information to related USGS science mission areas, especially the Climate and Land-Use Change and Ecosystems Mission Areas. Water-quality constituents to be monitored for characterizing surface water quality include:

- major ions,
- nutrients (N, P, and C),
- suspended sediment,
- pesticides,

- volatile organic compounds,[2]
- human and veterinary drugs,[3]
- semi-volatile organic chemicals,[4]
- algal toxins, and
- pathogens.

Water-quality constituents to be monitored for characterizing groundwater quality include:

- geochemical indicators,[5]
- age-dating tracers,
- major ions and nitrate,
- trace elements,
- pesticides,
- volatile organic compounds,
- human and veterinary drugs,
- semi-volatile organic chemicals,
- radionuclides,[6] and
- pathogens.

NAWQA has defined seven objectives for this goal; most of the Goal 1 objectives are viewed as central for NAWQA by the committee. However, status and trends networks consume more resources than other NAWQA activities, so cautionary advice is also included in the following discussion (Figure 2-10). Objective 1a is the long-term status and trends assessment of surface water, and Objective 1d is the long-term status and trends for groundwater. NAWQA cannot meet its core mission, let alone Cycle 3 Goals, without collecting these data. Objective 1g is the status and trends assessment of the biological condition of the nation's surface waters, which provides an assessment of water quality beyond what chemical measurements alone can provide. Relating biological condition to chemical and physical conditions can provide insight into likely factors causing degradation. Objective 1f is the assessment of contaminant loads to receiving waters. Given the importance of water-quality issues such as, for example,

---

[2] This constituent group includes disinfection byproducts, select high production volume chemicals, and volatile organic compounds.

[3] This constituent group includes antimicrobials, pharmaceuticals, and hormones.

[4] This constituent group covers a wide variety of trace organic chemicals, some occurring naturally but most associated with waste and wastewater. Chemicals include those found in detergents, flame retardants, and personal care products.

[5] These include basic properties such as temperature, pH, specific conductance, and other indicators of redox conditions such as dissolved oxygen.

[6] This constituent group includes uranium, radon, lead, and polonium.

Gulf Coast hypoxia, Objective 1f is considered essential. NAWQA stake-holders also expressed the essential nature of this objective at the 2009 meeting of the National Liaison Committee. Furthermore, the data collected in pursuit of this objective are the foundation for SPARROW.

For these essential Goal 1 objectives, NAWQA has identified informational needs that were not addressed during Cycles 1 and 2 and expanded each goal. An example is the addition of sediment, one of the four major water-quality stressors (Figure 4-1) to the long-term, national status and trends assessments of surface water (mentioned specifically in Objectives 1f and 1g). This includes adding sediment as a national synthesis assessment topic (NAWQA leadership, personal communication, October 26, 2010). The NRC (2002) presented a compelling discussion on the importance of conducting sediment monitoring, including suspended sediment, excess sedimentation, and particle-associated contaminants, and interpretation of this monitoring. National-scale sediment assessment was not pursued in Cycle 2 because of limited funding.[7] This committee encouraged the program to pursue sediment monitoring (NRC, 2010), noting it was a valuable scientific pursuit. However, given the magnitude of resources likely required to pursue sediment monitoring at the scale and detail proposed in the Science Plan, caution is advised. NAWQA would be well served by strategic investment in sediment monitoring, for example, through pursuit in select watersheds, choosing top priority topics related to sediment, and/or using SPARROW as a central tool.

Objective 1a, while representing the long-term status and trends surface water network, includes lakes and reservoirs. NAWQA, by design, does not sample many lakes or reservoirs, but it has been encouraged to probe these water bodies in the past (NRC, 1990, 2002). NAWQA has not followed these suggestions because of limited funding. In light of the current fiscal climate, the committee advises caution when considering this part of Objective 1a. The Great Lakes, for example, are considered coastal systems and so often fall under the jurisdiction of the National Oceanic and Atmospheric Administration (NOAA). In addition, under the Great Lakes Water Quality Agreement between the United States and Canada, EPA is the official party to address water-quality issues.[8] Other units of USGS are active in the Great Lakes, however, and there is a USGS Great Lakes Science Center in Ann Arbor, Michigan (part of the former Biology Division, now the Ecosystem Mission Area[9]). Many large water suppliers

---

[7] Smaller-scale activities and targeted monitoring studies for sediment were conducted in both Cycle 1 and Cycle 2; see Chapters 2 and 3.

[8] The Great Lakes Water Quality Agreement commits the United States and Canada "to restore and maintain the chemical, biological, and physical integrity of the Great Lakes Basin Ecosystem;" see http://www.epa.gov/glnpo/glwqa/1978/index.html.

[9] See http://www.glsc.usgs.gov/.

that take water from lakes or reservoirs have some data about their source water. However, data for small systems are much less likely to be available, although the national significance of data from small systems is questionable because contaminants in those water bodies, if present, likely come from local sources.

In Objective 1d the Science Plan proposes addressing spatial gaps in knowledge of principal aquifers and ancillary data to interpret some changes in water quality. The spatial gaps may indeed be critical to understanding the principal aquifers, while some of the ancillary data needs should be evaluated for cost-effectiveness and benefit to understanding water quality. Likewise, embedded in the objectives in Goal 1 is monitoring of contaminants of emerging concern. The committee echoes the cautionary advice from NRC (2010)—define scientific concerns with respect to monitoring these compounds and not get caught up in the "contaminant of the day." In summary, although many of these objectives in Goal 1 are "essential," NAWQA should understand the trade-offs associated with pursuing the newer components of this goal.

Goal 2 Stressor Effects: Objectives "a" (hydrologic factors), "b" (sources), "d" (susceptibility), and "e" (effectiveness of practices)

Goals 2 and 3 represent the planned extension of Cycle 3 into "understanding" water-quality status and trends, per the original program design (Cycle 1, status; Cycle 2, trend assessment; Cycle 3, understanding). Recall the aforementioned advice that, first and foremost, "the NAWQA program should continue the basic monitoring needed to maintain the national status and trends assessment." Understanding studies, while valuable, cannot be done without the basic status and trends monitoring. It is clear that the objectives within this goal are intimately tied together and almost need to be viewed as a unit when discussing trade-offs.

The committee considers most of the objectives in Goal 2 to be core to the program mission (Box 4-4). Objectives 2a, 2b, 2d, and 2e directly address the "understanding" component of NAWQA. They relate to how hydrologic systems as well as sources, transport, and fluxes of contaminants, nutrients, and sediment are affected by land use, climate, and natural factors. Specifically, Objective 2a can be called the "hydrology matters" objective, pursuing understanding of how hydrology impacts water quality. This is a concept that was originally pursued in NAWQA's Agrochemical Sources, Transport and Fate topic study in Cycle 2, and emphasizing this as a larger program objective coincides with research challenges and opportunities facing the field of hydrologic science at the nexus between hydrology and water quality (NRC, 2012). Hydrology is a basic area of expertise within NAWQA and USGS; linking hydrology to water quality is an important and logical undertaking for the program.

BOX 4-4
NAWQA Science Plan, Goal 2 and Objectives

Goal 2: Evaluate how human activities and natural factors, such as land use and climate change, are affecting the quality of surface water and groundwater.

Essential Objectives:
2a. Determine how hydrologic systems—including water budgets, flow paths, travel times and streamflow alterations—are affected by land use, water use, climate, and natural factors.

2b. Determine how sources, transport, and fluxes of contaminants, nutrients and sediment are affected by land use, hydrologic system characteristics, climate and natural factors.

2d. Apply understanding of how land use, climate, and natural factors affect water quality to determine the susceptibility of surface-water and groundwater resources to degradation.

2e. Evaluate how the effectiveness of current and historic management practices and policy is related to hydrologic systems, sources, transport and transformation processes.

Important but Not Essential Objectives:
2c. Determine how nutrient transport through streams and rivers is affected by stream ecosystem processes.

SOURCE: Design of Cycle 3 of the National Water-Quality Assessment Program, 2013-2023: Part 2: Science Plan for Improved Water-Quality Information and Management; January 28 2011.

Intimately linked with Objective 2a is Objective 2b, which deals with understanding how contaminants are tied to large drivers or parameters such as land-use change, climate change, and geology. The committee previously identified these as the important drivers for NAWQA to consider (NRC, 2010). Objective 2d deals with building the process-level understanding achieved in Objectives 2a and 2b into models—a "cause and effect" analysis. Recalling the previous advice, a focus of NAWQA efforts in Cycle 3, second only to basic monitoring activities, should be support of modeling initiatives. At its core, Objective 2e is the impact piece of Goal 2, simulating environmental scenarios and evaluating how different management practices and policy translate to water-quality changes. The outcomes and products from meeting these objectives will be a series of models, maps, and web-based tools that describe these complex relationships and communicate them in a clear and useful manner. They can be used to evaluate management choices and inform policy decisions; thus, the committee finds Objective 2e essential.

Goal 3 Multiple Stressors: Objectives "b" (nutrient levels that initiate impairment), "c" (sediment and impairment), and "d" (effects of stream flow alteration)

The essence of Goal 3 is examining the effects of water-quality parameters, what is causing changes in water-quality, and the relative influence of each stressor (Box 4-5). Like Goal 2, Goal 3 represents the "understanding" piece of the Cycle 3 design and should be viewed in the context of the earlier recommendation about status and trends monitoring. Also, the objectives in Goal 3 are intimately linked and should almost be viewed as a unit when evaluating trade-offs.

The committee views Objective 3b as a core objective. It addresses a water-quality problem of national significance, and the resulting products will have clear relevance for policy decisions (e.g., EPA or states to which EPA has given primacy establishing numeric nutrient criteria under the Clean Water Act). The work proposed is primarily intensive studies, but studies for this objective should also capitalize on the wealth of nutrient and biological data available from Cycles 1 and 2 and the work proposed under Goal 1 of the Science Plan. Although biological responses may not be directly related to observed nutrient concentrations, as the Science Plan argues, numeric nutrient criteria are based on concentrations; hence the work proposed for this objective will be more relevant to policy decisions if concentrations are an essential component of the analyses.

The committee considers Objective 3c to be a core objective despite minimal pursuit of this topic in the past because of funding constraints. Excess sediment is a nationally significant source of water-quality impairment, USGS has unique expertise to address this issue, and the understanding promoted from intensive studies will be used to develop a predictive ecological model that can be used to assess the impact of excess sedimentation at regional scales. Indeed, the NRC (2002) supported the inclusion of sediment in the NAWQA portfolio, as did the two previous letter reports from this committee (NRC, 2009, 2010).

The committee also views Objective 3d as a core objective. Streamflow has been considered a "master variable" in stream ecosystems (Poff et al., 1997), and anthropogenic alteration of streamflow is widespread with impacts on stream biota and ecosystem processes (Carlisle et al., 2011; H. John Heinz Center for Science, Economics, and the Environment, 2008).

BOX 4-5
NAWQA Science Plan, Goal 3 and Objectives

Goal 3: Determine the relative effects, mechanisms of activity, and management implications of multiple stressors in aquatic ecosystems.

Essential Objectives:
3b. Determine the levels of nutrient enrichment that initiate ecological impairment, what ecological properties are affected, and which environmental indicators best identify the effects of nutrient enrichment on aquatic ecosystems.

3c. Determine how changes to suspended and depositional sediment impair stream ecosystems, which ecological properties are affected, and what measures are most appropriate to identify impairment.

3d. Determine the effects of streamflow alteration on stream ecosystems and the physical and chemical mechanisms by which streamflow alteration causes degradation.

Important but Not Essential Objectives:
3a. Determine the effects of contaminants on degradation of stream ecosystems, which contaminants have the greatest effects in different environmental settings and seasons, and evaluate which measures of contaminant exposure are the most useful for assessing potential effects.

3e. Evaluate the relative influences of multiple stressors on stream ecosystems in different regions that are under varying land uses and management practices.

SOURCE: Design of Cycle 3 of the National Water-Quality Assessment Program, 2013-2023: Part 2: Science Plan for Improved Water-Quality Information and Management; January 28 2011.

Studies described under this objective not only are relevant to current issues of altered streamflow, but also will be an essential component of NAWQA's analyses of impacts of climate change on water quality and stream ecosystems. These studies capitalize on the extensive body of streamflow data unique to USGS and include collaboration with others working on this issue at different scales.

Goal 4 Future Predictions: Objectives "a" (evaluate suitability of existing models for future scenarios) and "b" (develop decision support tools)

The essence of Goal 4 is the extension of NAWQA modeling work using knowledge gained in pursuit of Goals 1-3 (Box 4-6). This is a major

THE WAY FORWARD FOR THE THIRD DECADE OF NAWQA

BOX 4-6
NAWQA Science Plan, Goal 4 and Objectives

Goal 4: Predict the effects of human activities, climate change, and management strategies on future water quality and ecosystem condition.

Essential Objectives:
4a. Evaluate the suitability of existing water-quality models and enhance as necessary for predicting the effects of changes in climate and land use on water quality and ecosystem conditions.

4b. Develop decision-support tools for managers, policy makers, and scientists to evaluate the effects of changes in climate and human activities on water quality and ecosystems at watershed, state, regional, and national scales.

Important but Not Essential Objectives:
4c. Predict the physical and chemical water-quality and ecosystem conditions expected to result from future changes in climate and land use for selected watersheds.

SOURCE: Design of Cycle 3 of the National Water Quality Assessment Program, 2013-2023: Part 2: Science Plan for Improved Water-Quality Information and Management; January 28 2011.

---

thrust of Cycle 3, which the committee supports. Objective 4a is essentially a study of how to enhance existing water-quality models. Without suitable models, the ability to gain understanding and forecast with greater precision is jeopardized. Models like SPARROW and WARP, two NAWQA mainstays, have proven themselves suitable but have to be enhanced to account for dynamic conditions and nonstationarity (Milly et al., 2008). Thus, Objective 4a is essential. Yet models that are as yet untested or even non-existent will likely be required. As an example of the former, a potentially significant quantitative tool for assessing ecological condition is the current effort to quantify the extent and severity of streamflow alteration (Carlisle et al., 2009); understand the relationships among land use, climate change, and streamflow alteration; and quantify relations between streamflow alteration and biological impairment (Carlisle et al., 2011). Time will tell if this approach has merit, but the work to date appears promising.

Objective 4b, calling for the development, calibration, and validation of decision-support tools, is essential to maintaining and enhancing NAWQA's policy relevance. Decision-support tools are essential to water quality and water resources management in general. USGS and NAWQA in particular have not been known for the development of decision-support tools, mainly because it is a new pursuit for the agency. NAWQA should identify clients for which it can develop and test decision support tools. Evolving SPAR-ROW to be a dynamic rather than a steady state model should be an aspect of this goal.

Objectives That Are Important but *Not Essential* to Cycle 3

The committee advises that the following objectives are important but not essential to NAWQA's mission and to its role as a national program. For these objectives, the committee believes that NAWQA should play a contributing role or work closely in partnership with other organizations with complementary capabilities (for additional information, see Chapter 5). These are identified in Boxes 4-3 through 4-6, above.

Goal 1 Status and Trends: Objective c (status and trends of microbial contaminants)

In Cycle 1, indicator bacteria were collected at study unit monitoring sites, but the data were never synthesized at the national level. At the end of Cycle 1, a pilot study was conducted to evaluate the occurrence of indicator organisms in stream and groundwater sites (Francy et al., 2000). This pilot was intended to support and inform larger-scale monitoring of microbial contaminants in Cycle 2, which while supported by NRC (2002), was not pursued because of limited funding.

Although Objective 1c (determine the distributions and trends in microbial contaminants in streams and rivers used for recreation) is a valuable scientific effort, it is not considered core to NAWQA. In NRC (2010), the committee questioned whether NAWQA's pursuit of microbial contaminants (then, articulated as a water-quality stressor in the Science Framework) was within the scope of its vision. The committee reiterates that concern here. Furthermore, assessing the status and trends of microbial contaminants at the scale proposed in the Science Plan is a formidable task. The committee questions whether NAWQA has the capacity to proceed with this objective; this could be a resource-intensive effort, and it may be inappropriate to proceed at the expense of core efforts, given limited funding.

Yet this objective would have consequences if not undertaken. The essence of this goal is a human health issue, the result of which would establish the quality of recreational waters. Not only is the societal benefit clear, but also assessing microbial contaminants can be a highly visible activity for the program, clearly demonstrating program impact. NAWQA needs to examine the costs and benefits of obtaining these data when determining whether to pursue this objective. States have also been monitoring microbial contaminants in streams and rivers used for recreation, with some of the resources for these activities coming from EPA; collaborative opportunities exist. The USGS Energy and Minerals, and Environmental Health Mission Area is a logical partner. Finally, microorganisms have a major impact on the many chemical constituents that are the focus of NAWQA's core monitoring. The cautionary advice regarding Objective 1c is not to be interpreted as suggesting that NAWQA ignore the role of microbes in biogeochemical processes, for example, the importance of denitrifying bacteria to nitrate levels.

Goal 2 Stressor Effects: Objective "c" (stream processes on nutrient transport)

Objective 2c is intended to determine how nutrient transport through streams and rivers is affected by stream ecosystem processes. This is a relatively specific objective, representing an extension of the Effects of Nutrient Enrichment in Stream Ecosystems topical study in Cycle 2. This is an important but not essential understanding to have, although it is similar to the processes considered of core importance (Objectives 2a and 2b) because it addresses the feedback loops of how ecosystems affect nutrients rather than how nutrients affect ecosystems. If this objective has the same importance as Objectives 2a and 2b and is central to designing NAWQA models, then it needs to be more clearly justified. Other programs are addressing similar objectives (e.g., the National Science Foundation's STRem Experimental Observatory Network or STREON,[10] academics), so this is one of the objectives that may be best addressed in conjunction with other programs in a leadership role.

Goal 3 Multiple Stressors: Objectives "a" (effects of contaminants on stream ecosystems) and "e" (multiple stressors in different regions)

The committee considers Objective 3a to be a secondary objective for NAWQA. The committee recognizes that streams are subjected to multiple stressors, which is an issue of national importance; however, the committee is concerned that the level of effort required to adequately address this problem could consume most of NAWQA's resources. The scale of studies being proposed is not adequate to assess this problem at a national scale, and the studies proposed are being done by other agencies (e.g., EPA)

---

[10] STREON sites are a subset of the National Science Foundation's National Ecological Observatory Network (NEON). The STREON experiment is designed to study nutrient dynamics in streams across the United States.

and academics. Furthermore, the toxicological laboratory studies proposed seem outside the core mission of NAWQA.

Objective 3e is worthwhile, but the committee considers it to be secondary for NAWQA. The use of structural equation modeling and Bayesian network analysis are innovative and appropriate approaches for trying to understand the relative importance of multiple stressors on aquatic ecosystems. One concern with this objective is the scale at which the studies are to be conducted. The Science Plan does not articulate how this work would be used to provide a national assessment.

Goal 4 Future Predictions: Objective "c" (predictions of water quality and ecosystem condition for specific watersheds)

Objective 4c is important to scientific understanding and to policy and decision-making, so a discussion of trade-offs and the need for partnerships is particularly relevant in this case. Predicting changes in water quality and ecosystem conditions in response to changes in climate and land use are relevant in Cycle 3 with its forecasting emphasis. Those issues are being addressed in the modeling conducted per Objective 4a. Hence, the ability of NAWQA to make progress in Objective 4c is dependent on the level of success achieved with Objective 4a and the scale at which the models are developed.

The essence of Objective 4c, while laudable, is ambitious. In the Science Plan, NAWQA has listed watersheds in which these studies would occur, and most importantly, identified specific partners with whom these studies proposed in Objective 4c would be conducted (Table 4-1). The Science Plans notes that "within each study area, the study will focus on a crucial issue which will be identified by one of NAWQA's partners." Furthermore, the Science Plan mentions the EPA, the National Oceanic and Atmospheric Administration, and the U.S. Department of Agriculture (USDA) as critical partners in all objectives in Goal 4. The committee agrees that working with partners in data-rich watersheds will be essential for accomplishing this objective. This objective was placed in this category because the committee acknowledged the need for partnerships and that NAWQA does not necessarily have to lead these efforts.

The Science Plan is not clear about what "ecosystem conditions" will be considered per this objective, although a hypothetical example is given in the text that outlines a model that forecasts nutrient transport to Chesapeake Bay under different climate and land use scenarios. That seems a realistic modeling objective, but forecasting other "ecosystem conditions" (e.g., macroinvertebrate populations, primary productivity) does not; hence the committee's ability to further evaluate this objective was limited by a lack of clarity in what is meant by "ecosystem conditions."

TABLE 4-1 Potential Study Areas and Primary Partnerships Proposed in Objective 4c of the Cycle 3 Science Plan

Basin study areas	Primary Partnerships
Chesapeake Bay	Chesapeake Bay Program
Mississippi River	Gulf of Mexico Watershed Nutrient Task Force, Louisiana Universities Marine Consortium
Delaware River	WaterSMART
Colorado River	WaterSMART
Apalachicola-Chattahoochee-Flint Rivers	WaterSMART
Great Lakes	Great Lakes Restoration Initiative, Great Lakes Commission

SOURCE: Design of Cycle 3 of the National Water-Quality Assessment Program, 2013-2023: Part 2: Science Plan for Improved Water-Quality Information and Management.

Objectives That Need *Further Justification* in Cycle 3

Finally, in a third category, the committee identified an objective for which sufficient information to determine their value to the nation and their place within NAWQA is lacking, particularly when compared to those objectives labeled as "core."

In the Cycle 2 Topical Study, Mercury in Stream Ecosystems, NAWQA provided data on availability of mercury in streams in targeted areas around the country and how mercury makes its way into fish and other organisms in stream ecosystems (Brigham et al., 2009; Chasar et al., 2009; Marvin-DiPasquale et al., 2009). Objective 1b (determine mercury trends in fish) proposes that NAWQA continue this work and expand the effort to capture long-term monitoring of mercury status and trends in fish. Yet it is not clear that NAWQA should expand this work to the scale proposed in Cycle 3.

Should NAWQA choose not to pursue Objective 1b in Cycle 3, there is a clear trade-off in terms of program impact. The Cycle 2 mercury work gained significant public attention; when it was released, the USGS Office of Communications commented on the 2009 mercury study and received 20,000 "tweets" in response and discussion. The public took a particular interest in understanding if fish were safe to eat. Figure 3-2 also shows the uptick in website use when the mercury work was released. Also, further understanding of water-column chemistry and mercury stream dynamics is a valuable scientific pursuit.

Many states collect and analyze fish tissue (Food and Drug Administration standard fillet) from water bodies, over time, to provide consumption advice. For example, the state of New York has been analyzing fish tissue for mercury since the 1960s, and from 1999 to 2008 it obtained mercury data for more than 12,000 fish; these data document trends in many New

York waters.[11] Many other states also collect mercury data.[12] If NAWQA does not undertake these activities, the states, other federal agencies, and possibly academia might provide data, and in some cases, significantly more data, than NAWQA.

It is essential that the evaluation of trade-offs continues as the Science Plan evolves and throughout Cycle 3. The discussion presented here has merely scratched the surface and provides only a high-level evaluation of science priorities, because the actual details of Cycle 3 will be developed as part of the Implementation Plan. **The NAWQA team should continue to evaluate what is essential for the program and why during Cycle 3, and use this evaluation to guide investments and effort.**

CYCLE 3 DESIGN ELEMENTS

The Statement of Task (bullet 4) reads:

Review strategic science and *implementation plans* for Cycle 3 for technical soundness and ability to meet stated objectives. [Emphasis added]

Although the Implementation Plan for Cycle 3 was not yet prepared at the time of this review, the Science Plan contained a preliminary discussion of how to implement the scientific agenda within. The preliminary design elements of Cycle 3 appear to be technically sound (NRC, 2010 and the discussion below). In the Science Plan, NAWQA proposes to increase coverage (i.e., increase the number of sampling sites) to better meet national needs assessment (Table 4-2, Cycle 3 (planned)). This increased coverage would bring the NAWQA sampling network closer to the number of sites proposed in the original design. But, the design elements for collecting data in Cycle 3 should also be cast in the context of the inevitable trade-offs that will occur to implement the program under current fiscal conditions.

Robust Sampling Plan for Status and Trends Monitoring in Cycle 3

The National Fixed Site Network (NFSN)[13] has been the core component of NAWQA through Cycles 1 and 2, underpinning status and trends

---

[11] The state of New York analyzed data from a New York State Department of Health comprehensive database of mercury levels in New York State sportfish (analyzed as standard fillets). The New York State Department of Health database compiles data sets provided by the New York State Department of Environmental Conservation, the New York State agency that monitors contaminant levels in fish.

[12] See http://water.epa.gov/scitech/swguidance/fishshellfish/fishadvisories/index.cfm.

[13] The NFSN is defined in the Science Plan as "a national network of monitoring sites that serves as the foundation for systematic tracking of the status and trends of stream and river water quality and for supporting and linking shorter-term studies at smaller scales." In Cycle 2, this network was referred to as the National Trend Networks.

analysis (primarily linked to Goal 1 objectives) and anchoring more detailed understanding assessments. In Cycle 1 there were approximately 500 NFSN sites, later reduced to approximately 140 sites and then to 113, often with decreased sampling frequency. The Science Plan proposes increasing the NFSN by monitoring sites every year (rather than a rotational schedule of intense monitoring every 2 to 4 years), real-time monitoring of select parameters such as turbidity, and additional sites to fulfill the expanded Cycle 3 goals (the Science Plan proposes increasing from 113 to 313 sites; Table 4-2).

The Science Plan justifies yearly monitoring using the example of diazinon concentrations in stream water responding to bans on the indoor and outdoor use of this pesticide. This provides one of many examples of how increased sampling has resulted in major insights into regulatory actions (Box 4-7), a justification that the committee finds compelling. In this case, the response to and assessment of a policy decision would not have been possible with samples taken at 2- or 4-year intervals. Furthermore, enhanced spatial coverage will facilitate ecological and climate change analysis. Continued status and trends assessment using sites near coastal areas will improve assessment of contaminant loads to hypoxic estuaries. The NFSN shares sites with other national programs, and Cycle 3 proposes to expand these collaborative efforts supported by the National Stream-Quality Accounting Network (NASQAN), the Hydrologic Benchmark Program, the Global Change Program, and the interagency National Monitoring Network. However, the expansion to coastal monitoring sites (sites situated farther into the coastal zone than those intended in Objective 1f) and, particularly, sites for drinking water source evaluation are not essential to NAWQA's core mission.

Although sampling the NFSN every year appears necessary to obtain an understanding of policy actions, the Science Plan does not provide adequate justification for adding the *number* of sampling sites that have been proposed, 313. Given the cuts that were made to the program in Cycle 2 to the point that the network is significantly reduced compared to original size, the committee believes that there is a need for additional sites. One indicator is that the SPARROW model was originally calibrated in 1992 with approximately 500 surface water sites that, at that time, were the combination of NAWQA and NASQAN surface water networks. In testimony to this committee, NAWQA scientists noted that calibration of the updated SPARROW modeling efforts is becoming more and more difficult because of the loss of sampling sites and the corresponding impact on model prediction error. (The use of other agency data is alleviating some of this difficulty, see Chapter 5.) However, a justification for the number of sites to be added and the criteria that will be used for choosing which sites to add are critical. Some analysis is needed of what would be gained by adding

TABLE 4-2 A Summary of NAWQA Program Design by Cycle Showing the Evolution of Program Design Since 1991

	Cycle 1 1991-2001	Cycle 2 2002-2004	2004-2007	2007-2012	Cycle 3 (planned) 2013-2023
Number of Study Units (SUs) or Integrated Watershed Studies (IWSs)	51 (SU)	42 (SU); transition to Major River Basins and Principal Aquifers beginning in 2004			20 (IWS)[a]
Number of Multi-year, Regional Assessments	n/a	n/a	8 Major River Basins; 19 Principal Aquifers	8 Major River Basins; 19 Principal Aquifers	8 Major River Basins, 24 Principal Aquifer Assessments
Number of Regional Synoptic Studies	n/a	n/a	n/a	n/a	10-20
Number of surface water sampling sites in Fixed Site Monitoring Network	505	145	84	113	313[b]
Sampling frequency of fixed surface water sampling sites	One-third of SUs sampled intensively every 3 years with 18-30 samples per site per year, only one-fourth of sites continued to be sampled after-intensive period ended	6-30 samples per year (most sites sampled 8 times per year), all years	6-26 samples per year (most sites sampled 6 times per year), all years	6-26 samples per year (most sites sampled 16 or more times per year), with most sites monitored 1 out of every 4 years	18-24 samples per year, all years

Number of aquatic ecology sites[c]	416	125	75	58 (6 sites are ecology-only)	88
Sampling frequency of aquatic ecology sites	At least once; subset of sites were sampled annually during 3-year high-intensity phase	Annually, beginning 2002	Annually (biennially for fish), 2005-2006	Every 2 years (invertebrates and algae annually at reference sites)	Annually
Number of groundwater networks/wells	272 networks, 6,307 wells	137 networks, 3,698 wells			170 networks, 6,450 wells
Additional studies	High Plains Aquifer study	Topical Studies; Source Water Quality Assessments			Regional Synoptic Studies; Intensive Studies;[d] Regional Groundwater Studies;[e] Local Groundwater Studies[f]

[a] IWS can be considered surface water focused "study units" where an emphasis on understanding hydrologic linkages bewteen contaminant sources and transport both in surface water and groundwater are studied. NAWQA plans for 1-2 IWS in each Cycle 2 Major River Basin. The IWS will consist of core assessment activities but will also be customized to address location conditions.

[b] Includes 70 drinking-water intakes, with 20 on streams and 50 on reservoirs.

[c] The ecology sites are included in the total number of surface water sampling sites.

[d] Regional Synoptic Studies are short-term, targeted water-quality assessments of specific regional and (or) water-qulality conditions that generally overlie one or more IWS areas. Intensive Studies are interdisciplinary studies ranging in scale from individual stream reaches to small watersheds and are planned to be nested within the IWS. Both are surface-water focused.

[e] Regional Groundwater Studies are nested within Principal Aquifers and designed to contribute to assessment of status and trends at the regional to national scale and also, by the use of regional flow models, insights into regional groundwater contributions of water and contaminants to streams.

[f] Local Groundwater Studies mimic Cycle 1 and 2 Flow System Studies and are designed to improve understanding of groundwater quality at a more specific, local flow-path scale. The Intensive Studies and Local Groundwater Studies will be co-located and nested within Regional Groundwater Studies and the IWS and are intended to provide insights regarding surface water and groundwater interactions.

BOX 4-7
The Importance of Increased Sampling

Beginning in the mid-1990s, NAWQA collected samples and probed the presence of the insecticide diazinon in an urban stream. Samples were collected annually, rather than on the 4-year rotational sampling design commonly employed by NAWQA during Cycle 2. NAWQA continued sampling as diazinon was phased out for both indoor and outdoor residential use in the early 2000s, and developed a reliable time-series model to assess long-term changes in diazinon concentrations as residential use declined. The model showed a rapid water-quality response to eliminating outdoor uses in 2002 and a continued decline in diazinon concentration through 2004. NAWQA then reanalyzed the same data using only the information that would have been available if the 4-year rotational sampling design had been employed, i.e., if the model was based on sampling every fourthyear. The resulting trend indicated an *increase* in diazinon in streams through 2004, rather than the decrease in concentration that had actually occurred. If NAWQA had not sampled annually, then the effectiveness and environmental benefits of the regulatory decision to phase out diazinon would have been called into question.

SOURCE: Modified from NRC, 2010.

---

different numbers or some combinations of sites, with clear explanation of the criteria used for those choices.

Furthermore, with the basic study design changes since Cycle 1, the 313 sites used under the proposed study design may not be the most appropriate for the design and objectives of Cycle 3. The committee has a similar concern regarding groundwater sampling design; there is insufficient information to evaluate whether the number of sites (3,000 monitoring wells, 2,500 domestic wells, and 700 public wells) is too few or too many to meet the Cycle 3 objectives. NAWQA is correctly mindful of maintaining sites where long-term trend data have been collected, and this commentary is not to be interpreted as discontinuing these valuable sites. However, given the planned emphasis on modeling in Cycle 3 it is important that the design corresponds to this emphasis.

NAWQA used both a linear programming approach and an expert judgment based on semiquantitative analysis to select the reduced number of study units at the beginning of Cycle 2. When used in conjunction, these approaches ensured that the Cycle 2 status and trends network would account for at least 50 percent of the nation's drinking water use,[14] a cross-

---

[14] When completed, the final group of study units accounted for 61 percent of the national drinking water use.

section of the nation's hydrologic settings and ecological regions, the top 10 regions representing major contaminant sources (urban, agriculture, and natural), and major aquifer systems. The study units were prioritized based on these criteria, and those not evaluated as top priority were revisited to ensure that they did not possess characteristics that would warrant their inclusion in the priority list. These approaches are discussed extensively in NRC (2002); the 2002 NRC committee concluded that these approaches were "commendable."

Following a similar path, NAWQA's Surface Water Status and Trends Redesign Committee was created in the mid-2000s to modify the Cycle 2 design and operation of networks because of concerns about rising program costs in an environment of stable or declining appropriations. In making recommendations for the redesign, the committee considered the fiscal environment, scientific evidence, and maintenance of established sites with a relatively long trend record, all within the framework of remaining true to the original objectives of the program. The redesign committee made two major recommendations on which NAWQA acted: (1) the program should take full advantage of the use of models to define agricultural status and trends and answer large-scale questions by extrapolation and (2) the program should emphasize the national and regional scales through Major River Basins (NAWQA leadership, personal communication, March 19, 2012).

These efforts and NAWQA's flexibility during Cycle 2 are commendable, and they can be of use to the program both in the implementation of Cycle 3 and if the program faces similar challenges in the future. **NAWQA should determine the number of sampling locations and frequency using a similar process that was used in Cycle 2, adapted to the objectives for Cycle 3, with particular consideration of the certainty required for Cycle 3 modeling efforts.** This approach can aid an explicit determination of the budgetary implications of these decisions and options. This is likely the purview of the forthcoming Implementation Plan for Cycle 3, and this advice is to be taken in this context.

In the second letter report, the committee recommended that the NAWQA monitoring and modeling design should reflect a dynamic sampling strategy, overlain on top of a periodic sampling design. By "dynamic," the committee means a design that is flexible to capture specific events (such as spring melt, or the first few inches of rainfall as runoff) or geographic scales (intensive sampling in a targeted area to capture a specific process). Such flexibility might be needed to provide optimal data for model calibration and validation, or to reduce uncertainty in certain model processes. Because monitoring is expensive, dynamic sampling should be used judiciously and where it will best reduce uncertainty in outputs. This monitoring may be done with collaborators (states, academics, etc.), taking full advantage of real-time measurement technologies.

NAWQA has always used a nested hierarchy of sites, or design elements, in both surface water and groundwater studies to enable spatial and temporal extrapolation. For surface waters, this nesting involves locating smaller watersheds within larger watersheds at different scales by sharing key sites. Nesting groundwater sites spatially and at different depths contributes to a three-dimensional understanding and permits spatial extrapolation. The design proposed in the Science Plan for Cycle 3 is no exception. Another basic design element used by NAWQA is retrospective analysis, compiling historical data, assessments, and insights gained from these analyses. Given its success, **NAWQA should continue using nested sites and retrospective analyses of program data, and also data from federal, state, and local partners, to maximize the coverage of their assessments.**

Regional Synoptic Studies (RSS) are targeted to address spatial gaps related to contaminant status and trends. **The use and addition of RSS sites should be closely evaluated with respect to their necessity in answering regional and national questions and their contribution to model development.** Use of sites maintained by other agencies and academic organizations should be explored because such collaboration could help reduce resource requirements and/or enhance the utility of NAWQA data.

Integrated Watershed Studies (IWS) are long-term water-quality assessments and are typically anchored by one or more NFSN sites. These sites are similar to the former study units in concept and represent the reincarnation of these former building blocks in the Science Plan. The committee supports the IWS but recognizes that pursuit might be limited to a pilot phase in the challenging fiscal climate. **Potential IWS should be closely evaluated to ensure that the sites selected will clearly contribute to solving regional and national questions and/or meeting key model development needs.** Some IWS may be well-suited for developing collaborative support with local, state, and federal agencies, such as EPA, and the U.S. Department of Agriculture, and perhaps even academic research teams.

Intensive Studies (IS) focus on individual small-scale watersheds (or even stream reaches) to address details of hydrologic and/or biogeochemical processes. Because of their scale, IS sites are ideally suited for developing collaborative interactions with local, state, and federal agencies, and academic research teams. NAWQA could very well defer these sites to others and collaboratively use their data. Such sites could, for example, be operated by groups such as the Toxics Substances Hydrology Program, the Global Change Program, or non-USGS programs like the National Science Foundation's National Ecological Observatory Network (NEON).[15]

---

[15] The National Science Foundation's National Ecological Observatory Network is a research instrument consisting of infrastructure distributed across the United States designed to conduct continental-scale ecological research.

With respect to groundwater, Cycles 1 and 2 focused on shallow groundwater, or younger, recently recharged waters. Cycle 3 proposes to build on this assessment and add further observations of deeper groundwaters within Principal Aquifers with the focus being on drinking water. The Principal Aquifer Assessments will be the primary unit for groundwater studies in Cycle 3 to assess the status and trends of groundwater on a national scale. Regional Groundwater Studies (RGS), nested within a Principal Aquifer, will be collocated with IWS surface-water studies. The third proposed groundwater design element, Local Groundwater Studies, will be nested within RGS and/or may be collocated with surface-water IS, to improve knowledge on specific cause and effect to increase understanding of human activities and natural processes that affect groundwater quality. The committee's advice to the program regarding mindfulness of the linkage between surface water and groundwater is consistent with this nested design.

The groundwater studies particularly depend upon collaborative efforts with the USGS Groundwater Resources Program, the Water Cooperative Program, the Cooperative Geologic Mapping program (of USGS and their state partners), and other federal and non-federal partners both for data acquisition and modeling input data. **These groundwater design elements and the addition of new sites should be carefully evaluated for their contribution to answering regional and national questions at NAWQA's core, and for their contribution to key model development needs, rather than focusing on more local-scale evaluations.**

COMMUNICATION AND PROGRAM IMPACT

NAWQA has used a wide array of approaches to communicate findings, from press releases to congressional briefings, peer-reviewed publications, and the program website. These efforts are directed by the NAWQA Communications Coordinator, an important role within the program. The committee has noted that these efforts are an accomplishment of the program (Chapter 3, Appendix C). Yet several communication challenges exist and are discussed below.

NAWQA has performed three Customer Satisfaction Surveys in the past 12 years, and each has had a slightly different format.[16] In the two earlier surveys NAWQA learned that users favor downloadable graphics; as a result the 2006 Pesticide Circular invested significant resources into developing downloadable graphics. NAWQA also discovered that 50 percent of their users are "technical" or use the models and more technical

---

[16] The first Customer Satisfaction Survey was in 2000, probing the usefulness of a specific product, *The Quality of Our Nation's Waters—Nutrients and Pesticides* (USGS, 1999). The second and third were both of a more general format and were conducted in 2004 and 2010.

components of the program's output. And of the overall audience, 80 percent found the fact sheets (which are geared to the non-technical audience; see Appendix C) useful, meaning the NAWQA fact sheets resonate with a broader audience than originally thought.

Results from the 2010 survey indicated that 90 percent of NAWQA stakeholders found the use of "email blasts" and the NAWQA website (in terms of navigation and relevance) effective. Yet, the majority of NAWQA stakeholders access the website only occasionally. NAWQA stakeholders identified sediment and contaminants of emerging concern as the two biggest information gaps in the program. The majority (95 percent) of stakeholders prefer electronic copies of NAWQA program documents. NAWQA program video casts (CoreCasts) are the least used product, and few respondents were interested in social media tools such as Facebook and Twitter. Yet audience response indicates that if results are easy to understand, video podcasts are an effective means of presenting scientific information (Moorman et al., 2011). Most respondents do not use the NAWQA data warehouse (P. Hamilton, personal communication, October 26, 2010). NAWQA stakeholders indicated satisfaction with the program website, but the majority visit the site only "occasionally" (Figure 4-2).

NAWQA does, informally, measure success and feedback on a more frequent basis than Customer Satisfaction Surveys. This includes monitoring the number of website hits, the number of requests for products at the time of release, and attendance at briefings during product launches, and collecting media coverage. However, this tracking is sporadic and lacks a structured approach and cataloging system. Thus, an opportunity exists. Working collaboratively and taking full advantage of expertise in the USGS Office of Public Affairs, **NAWQA should establish a formal mechanism to evaluate the success and effectiveness of all the elements in its public relations portfolio and adapt public relations efforts as needed.** This would be helpful not only in directing the communication efforts of the program but also in tracking and illustrating the importance of NAWQA. Indeed, the 2010 Customer Satisfaction Survey indicated that approximately 45 percent of those accessing NAWQA data use it in *policy development* (Figure 4-3). This is a critical piece of information, and NAWQA needs to know more.[17] Also shown by this survey was that some users of NAWQA information are dissatisfied with that information (Figure 2-14). Additional insights into the reasons for the dissatisfaction would be useful to the program.

---

[17] NAWQA's Customer Satisfaction Survey in 2000 showed a similar result (46 percent of NAWQA stakeholders used NAWQA information for "policy development and decision-making") but, given the differences in the two surveys, the committee is reluctant to compare results from the two.

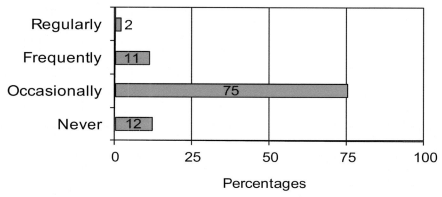

Percentages

FIGURE 4-2 The 2010 Customer Satisfaction Survey indicated NAWQA users "occasionally" visit the NAWQA website. SOURCE: USGS, personal communication.

Beyond formal tracking of communication efforts, multiple tools are needed to capture the impact of NAWQA products and information. For example, this might include a quantitative bibliometric analysis of publications or a formal assessment of website access and downloads, building on the information in Figure 3-1 of this report. Currently, one of the primary mechanisms for tracking program impact is a document titled *The National Water-Quality Assessment Program—Science to Policy and Management*. This document is available to the public through a live link on the NAWQA website's home page and is frequently updated by NAWQA personnel. Testimony in the document is a valuable indicator of program impact; indeed, information from the document is sprinkled throughout this report. However, **NAWQA should further highlight this document or the information contained therein, perhaps with a designated web page on program impact, to emphasize the value of NAWQA information to a variety of users.** An opportunity also exists to dovetail this type of information with that gleaned from a more formal mechanism in order to track the success and impact of the program.

In presentations to the committee, NAWQA leadership indicated a continued commitment to the NAWQA website, specifically promoting its availability such as through product-releases and links to related websites and organizations. This is a commitment the committee supports. The committee also supports continued development of innovative web-based dissemination tools such as video podcasts. Although the 2010 survey indicates they are not widely viewed, video podcasts are a new tool and the survey is only one assessment of their success. A more formalized mechanism for tracking the NAWQA public relationship portfolio will further

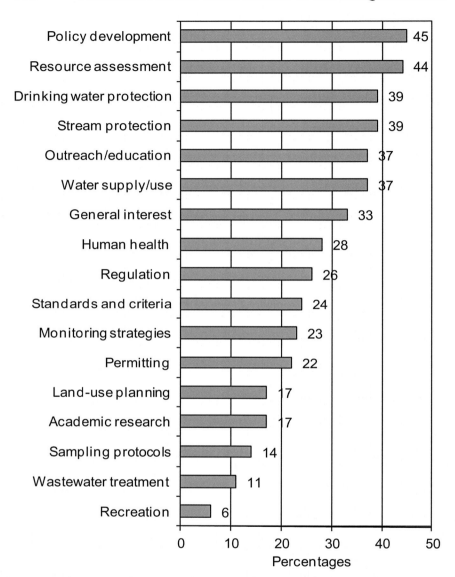

FIGURE 4-3 The 2010 Customer Satisfaction Survey indicated 45 percent of users use NAWQA information for policy development. SOURCE: USGS, personal communication.

determine the utility of these and other web-based efforts. Given the explosion of social media, the sense of the committee is that video podcasts are worth pursuing, when appropriate.

The committee acknowledged NAWQA's data warehouse and other tools to bring raw water-quality data to the public as an accomplishment in Chapter 3. However, the data warehouse is not nearly as user-friendly as, for example, the SPARROW Decision Support System interface (Box 4-1). Perhaps this is the reason that the Customer Satisfaction Survey respondents rarely use the data warehouse. The volume of data and the associated supporting data and metadata continue to expand exponentially, and NAWQA needs to ensure that it has a process for keeping up with these data and providing them to users (within and outside the agency) in a coherent manner. The data warehouse interface design should be evaluated and improved, with significant user input as to what should be included and how it should be presented. It will also need constant updating and adjusting. **Although the committee considers development of the data warehouse to be an accomplishment, further efforts to improve the data warehouse interface are needed.**

NAWQA should also look for innovative ways to ensure that data interpretation, synthesis, and publication take place in a timely manner. The committee acknowledges the difficulty of this task given the sheer size of the datasets that NAWQA scientists publish, the intense yet valuable USGS peer-review process, and resource constraints. Suggestions include the use of postdoctoral scientists, internship students, interagency collaborators, or the addition of staff dedicated to this endeavor. Perhaps increasing the availability of NAWQA data through the Internet would suffice, while the more time-intensive efforts (i.e., interpretation, synthesis, and publication) continue. Timely interpretation, synthesis, and release of NAWQA results is critical. **NAWQA data used in these results should continue to be delivered to the public via an improved public database.**

The committee believes it is critical to identify and document the cases where NAWQA data and analysis have influenced policy and decision making. Ultimately, tracking impact will allow NAWQA to demonstrate significance and the return on the nation's investment. Making a slice of this information available to the public could attract new users. **A unified strategy for the timely preparation, release, and subsequent tracking of the impact of NAWQA information and products is needed.** The committee realizes an effort such as this will require resources during a time when resources are stretched thin and encourages the use of the USGS Office of Public Affairs, when appropriate. The benefit of this exercise will far outweigh the associated challenges.

CONSEQUENCES OF PRIORITIZATION OF OBJECTIVES

In an ideal world, there would be sufficient resources to implement the Cycle 3 Science Plan. Recognizing that some objectives are more directly related to NAWQA's core functions than others, the committee believed it important to parse the Science Plan into what components are essential, need further justification, and are important but not essential to NAWQA's overall mission and goals. There are consequences of failing to implement the Science Plan in its entirety. For example, some of the activities surrounding the important but not essential objectives could be viewed as more policy relevant, intellectually challenging, and professionally satisfying than many of the activities associated with the essential objectives. This may have consequences as to the quality, productivity, and morale of the workforce. The impact of not studying all the process-oriented objectives in Goals 2 and 3 may limit the full development and accuracy of the models, because those processes may not be well characterized by current models. Objectives of least importance are those that could be addressed by others (states, academia, etc.) or are very regional in scale. Not addressing these objectives could mean they are never addressed.

Finally, the committee recognizes that this Science Plan and NAWQA itself will continue to adapt to change and some of the objectives could be phased in or addressed later in the decade. Other opportunities may arise to implement objectives through innovative collaborations with the many partners within USGS, the federal government, states, academe, and nongovernmental organizations. **Given the likelihood that NAWQA will have insufficient funding to proceed with the full scope of the Cycle 3 Science Plan, these opportunities should be actively identified and pursued.**

5
Coordination, Cooperation, and Collaboration

In the Science Plan for Cycle 3, the National Water-Quality Assessment (NAWQA) program put forth an ambitious strategy for continuing to monitor and assess the nation's freshwater quality and aquatic ecosystems. This plan was a product of NAWQA's two decades of experience and input from more than 50 stakeholder groups and partners, as well as the National Research Council (NRC). In contrast to Cycles 1 and 2, the Science Plan for Cycle 3 offers a vision that extends beyond NAWQA's organizational capacity and resources. It offers a vision for the nation and a strategy to address many key national needs, both for and by the many agencies and organizations concerned with water quality—not just NAWQA. This is a vision that the committee strongly supports.

Given the broad scope of the plan and the budget constraints already evident, cooperative, coordinated, and collaborative efforts should play a much greater role to meet many of the needs identified in the Science Plan. (Definitions of these terms as they apply to NAWQA are clarified in Box 5-1.) Although NAWQA should be a leader in motivating the implementation of this national plan, other groups and agencies will need to play leadership roles to accomplish many of the goals and objectives in the Science Plan. The Science Plan for Cycle 3 is a plan for addressing national water-quality needs that deliberately goes beyond what NAWQA can accomplish. The Science Plan could be a framework for other agencies to identify objectives that they can meet as part of their own mission and at the same time support a larger and collective effort to address the nation's water-quality problems.

BOX 5-1
The Use of the Terms Cooperation, Coordination,
and Collaboration in this Report

Cooperation, coordination, and collaboration are complementary but distinct activities within the context of the NAWQA program.

- Cooperation: Sharing goals, plans, data, and other information within USGS and with other federal, state, and local agencies as well as stakeholder groups to increase awareness and reduce inter- and intra-agency friction.
- Coordination: In addition to the activities captured in the definition of cooperation, the committee uses "coordination" to mean proactive efforts by NAWQA to work within USGS and with other agencies and partners to ensure compatibility of goals, data gathering, and other program activities. If done well, coordination increases programmatic efficiency and reduces redundancies and conflicts.
- Collaboration: Taking coordination one step further, collaboration implies working together to conceptualize, plan, fund, and implement activities that lead to a larger understanding and programmatic impact that could not have been achieved if NAWQA and its partners acted independently.

The development of the comprehensive Science Plan with the input of the NAWQA National Liaison Committee (NLC), U.S. Geological Survey (USGS) personnel, other stakeholders, and this committee (NRC, 2010) is a clear example of the effort NAWQA has successfully put forth to work toward a cooperative, coordinated, and collaborative program. The committee commends NAWQA for its work in this arena and concurs with past reviews of NAWQA (NRC, 2002) and USGS Water Programs (NRC, 2009) that have cited NAWQA as exemplary for its efforts to establish cooperative relationships within USGS and with external stakeholders. There are many examples of such cooperative efforts throughout this report. Such cooperative efforts contribute to program and policy relevance and provide additional opportunities to communicate NAWQA's broader message of leveraging activities of others.

To successfully implement the Cycle 3 Science Plan, NAWQA will need to place even greater emphasis on collaborative efforts in which it is already engaged. These involve data sharing, interpretive efforts, and even mutual planning. However, NAWQA in Cycle 3 will need to go beyond these existing efforts to establish more active collaboration with external agencies and organizations (e.g., related to budgets and staffing) in which NAWQA and these partners work toward common assessment and other scientific goals. The effort will require a change in approach for parts of NAWQA in order to more fully and directly involve these potential partners and collaborators in the development of science and implementation workplans and budgets, explicitly outlining roles, responsibilities, and accountability.

In this chapter the committee draws out key points related to NAWQA's cooperative efforts to respond to the Statement of Task:

> Identify and assess opportunities for the NAWQA Program to better collaborate with other federal, state, and local government, non-governmental organizations, private industry, and academic stakeholders to assess the nation's current and emerging water quality issues.

The committee also identifies current and continuing challenges noted in the testimony from various agencies, identified in its deliberations, or heard from NAWQA's leadership.

Although cooperation, coordination, and collaboration are critical to meeting the goals of Cycle 3, the committee recognizes that these efforts are not as simple as they sound and indeed can be costly and time-consuming when trying to maintain communications among different parties. Difficulties can often arise from overlap or differences in missions that require management time to reconcile. Given resource constraints, partnering with other agencies will almost always be in NAWQA's interest if at least two conditions are met: (1) the contributions of the entity are methodologically consistent with NAWQA's analytical standards or some adjustment can be made to account for the lack thereof and (2) the relationship is likely to expand the reach and impact of the program. Obviously, if the transaction costs associated with such partnerships exceeds the financial benefits to the program, NAWQA would be better off declining the opportunity. Keeping this in mind, the committee makes a case for such efforts, seeing value in NAWQA's ability to leverage greater resources and expertise from external partners to meet the nation's needs for water-quality assessment and understanding.

NAWQA'S VALUE IN A REORGANIZED USGS: COOPERATION, COORDINATION, AND COLLABORATION WITH USGS MISSION AREAS AND PROGRAMS

NAWQA's scope and success providing a national perspective on the status, trends, and understanding of factors that affect water quality have made the program a visible and respected focal point within the Water Mission Area of the USGS.[1] As noted in past reviews (NRC, 2002), many

---

[1] Other programs and activities within the Water Mission Area include the Groundwater Resources Program, the National Streamflow Information Program, Hydrologic Research and Development, Hydrologic Networks and Analysis, the Cooperative Water Program, and the Water Resources Research Program.

local, state, and even federal agencies and organizations that had not worked with USGS in the past now regularly promote the use of USGS products and information because of their involvement with NAWQA. As one of the largest water programs within USGS, NAWQA has worked at cooperative efforts within USGS and the Department of the Interior (DOI) since the beginning of the program. Again, past reviews have generally commended these efforts, as well as pointed to areas for improvement (NRC, 2002, 2009). In particular, many reviews have applauded USGS for productive, collaborative symbiosis among field monitoring and research programs such as NAWQA, the National Research Program, and the Toxic Substances Hydrology ("Toxics") Program. These collaborations have made valuable contributions to the nation in areas such as contaminants of emerging concern and the development and broad implementation of the SPAtially Referenced Regressions on Watershed Attributes (SPARROW) model, as notable examples.

During the course of the committee's deliberations, and during the time the Science Plan was under development, USGS reorganized to enhance the work of the agency's science programs. The agency has historically been organized around technical disciplines (e.g., Biology, Geography, Geology, and Hydrology/Water) but has now aligned its leadership and budget structure around interdisciplinary themes or mission areas related to the science strategy "*Facing Tomorrow's Challenges—U.S. Geological Survey in the Decade 2007-2017*" (UGSG, 2007). The new mission areas are Ecosystems; Climate and Land-Use Change; Energy and Minerals, and Environmental Health; Natural Hazards; Core Science Systems; and Water. The realignment also created a new Office of Science Quality and Integrity tasked with monitoring and enhancing the quality of USGS science. The 2009 NRC report, *Towards a Sustainable and Secure Water Future,* pointed out that critical water-related issues occur within most if not all new USGS Science Strategy directions (now mission areas, in the official reorganization) (NRC, 2009). The report noted that approaching these new strategic directions will demand even greater coordination and cooperative efforts throughout USGS.

This committee's second letter report (Appendix B) provided some initial comments on NAWQA's possible place in the reorganized USGS. In that initial letter report the committee made comments and recommendations whose main concepts are reiterated here (Box 5-2). Water is now a theme running through several mission areas apart from the Water Mission Area itself—for example, Ecosystems and Climate and Land Use Change.

NAWQA leaders should seek further opportunities for cooperation, coordination, and collaboration within USGS and make a systematic effort to communicate its capabilities and potential value to the relevant programs and offices within USGS through the Science Plan. Also, during the time

---

BOX 5-2
Excerpts from NRC (2010)

"To enhance the work of the agency, the USGS is currently realigning its leadership and budget structure around interdisciplinary themes or mission areas related to the science strategy 'Facing Tomorrow's Challenges—U.S. Geological Survey in the Decade 2007-2017' (UGSG, 2007). . . . NAWQA is well positioned to contribute to these mission areas, building on its success in multidisciplinary efforts within the USGS over the last few decades . . ., but this is not well articulated in the [draft] Science Plan."

The letter report provided some specific notes:

"A continued relationship between NAWQA and programs in the Ecosystems Mission Area would be valuable to the USGS. NAWQA has integrated ecological components with physical and chemical measurements with the co-location of ecological and water quality sampling sites (NRC, 2009). NAWQA science has enhanced understanding of the effects of urbanization, mercury, and nutrients on stream ecosystems through Topical Studies in Cycle 2. NAWQA is currently developing a 'data warehouse' for biological information, in collaboration with other disciplines and programs within the USGS."

"NAWQA and the Toxic Substances Hydrology program (now part of the Energy and Minerals, and Environmental Health Mission Areas) have a long history of successful, joint collaboration (NRC, 2009; NRC, 2002). The USGS leads the way in identification, tracking, and doing research on emerging contaminants, a role resulting in part from collaboration between the USGS Toxic Substances Hydrology Program and NAWQA (Kolpin et al., 2002). . . ."

"One of NAWQA's noted accomplishments has been the linkage of land-use to water quality conditions. In Cycle 3, NAWQA proposes enhancing its consideration of climate change issues and water. This could be particularly valuable to and invite important collaborative opportunities with the Climate and Land-Use Change Mission Area. And certainly, NAWQA's long-standing work in data integration, as well as its experience developing a data warehouse to provide accessible data to other agencies and the public, is relevant to the work of the Core Science Systems mission."

The committee further noted:

"NAWQA has a history of working in the multidisciplinary, collaborative interface and could serve as a useful resource and model to assist in the realignment of the agency to multidisciplinary and cross-disciplinary missions. Although defining collaboration and listing partners is important to NAWQA planning efforts, true collaboration begins with identifying common questions or goals shared with other mission areas and USGS programs. **To be effective in this effort the Cycle 3 Science Plan, NAWQA should more clearly identify how its goals are linked to the newly formed USGS mission areas framed from themes in the USGS Science Strategy.**"

of the committee's deliberations, the WATERSmart program (an effort formerly referred to as the Water Census under the auspices of USGS; see NRC, 2009) was elevated to a DOI initiative. The WATERSmart program has been proposed to address the critical national need for water availability information, and its elevation may mean that it will receive high-level support for the collaborative effort needed to engage many of the other DOI bureaus. This effort will also require considerable cooperation with other federal, state, and local agencies to be successful (National Science and Technology Council, 2007; NRC, 2009). Water availability links water quantity and quality, and NAWQA will obviously be affected by the development and integration of the WATERSmart effort within DOI. NAWQA can be particularly effective in contributing to forecasts of water availability through the program's ability to relate its assessment of water quality and ecosystem health to changes in land use and land cover, natural and engineered infrastructure, water use, and climate change. Accomplishing this task effectively will require extensive interaction among scientists in the various USGS mission areas. NAWQA could add significant value to federal programs such as WATERSmart where there are important opportunities to address the national need for water-quality information.

During this committee's deliberations, the newly formed mission areas were developing strategic science plans and implementation plans. Organizational change always creates some disjointedness and dislocations during a transition phase, but the committee is concerned that in the tension of the transition phase, emerging goals for each Mission Area and competition for recognition and resources within USGS and DOI could be temporarily problematic for both NAWQA and the agency at large. For example, the Toxics program, one of NAWQA's closest collaborators, is now being housed in a separate Mission Area. Although the committee's second letter report challenged NAWQA leadership to communicate capabilities to the reorganized programs and to seek collaborative opportunities that would help meet the needs of the Science Plan that go beyond NAWQA, it also noted that such communication should be a two-way street. The committee would hope that USGS uses the reorganization to improve internal coordination and potentially leverage NAWQA data and analysis for use in the other program areas. Furthermore, fiscal realities highlight the need to seize these collaborative opportunities within USGS to make the most of these existing resources; the reorganization is a window of opportunity for this to be fully realized. Integration among the new mission areas presents important opportunities to leverage NAWQA activities for the benefit of USGS, the federal government, and the nation.

COORDINATION AND COOPERATION EFFORTS: NAWQA LIAISON COMMITTEES

Starting with its pilot studies, NAWQA began a particularly successful component with the development of local and national coordination and advisory groups. In establishing individual study unit liaison committees as a key component of NAWQA, USGS recognized the importance of relationship building and obtaining local information and perspectives on water-quality and water resource issues. These efforts fostered various partnerships and activities including local and state use of NAWQA data and SPARROW for Total Maximum Daily Loads (TMDLs); some local, jointly-funded projects (various projects on Source Water Assessments to protect public drinking water systems, such as Vowinkel et al. [1996], Ryker and Williamson [1996], and see USGS [2001]); and other federal efforts (NRC, 2002, 2009).

NAWQA's National Liaison Committee (NLC)[2] provides an ongoing platform for stakeholders to interact with NAWQA. Its purpose is three-fold, to:

1) exchange information on findings and about water-resource issues of national and regional interest, 2) identify sources of data and information, and 3) provide feedback on any Program changes, design, and scope of products.[3]

The committee is composed of approximately 100 participants spanning multiple federal agencies (U.S. Environmental Protection Agency [EPA], National Oceanic and Atmospheric Administration [NOAA], U.S. Department of Agriculture [USDA], Centers for Disease Control and Prevention, the Congressional Research Service, Department of Energy, etc.) and interested groups (American Water Works Association, Association of Metropolitan Water Agencies, American Rivers, National Association of City and County Health Organizations, Natural Resources Defense Council, etc.).[4] The NLC and many of its members offered input during the development of the Science Framework and the Science Plan to ensure that it provided comprehensive national relevance and appropriately captured stakeholder needs (Box 5-3).

The 2002 NRC review of NAWQA recommended that the local and national liaison committees should be continued in Cycle 2 and noted that

---

[2] See http://acwi.gov/nawqa/index.html.

[3] See http://acwi.gov/nawqa/.

[4] The NAWQA National Advisory Committee went through various iterations and restructuring, partly under the Federal Advisory Committee Act (FACA) requirements (see NRC, 2002), to reach its current structure as the National Liaison Committee for NAWQA. It has been formalized under FACA as a subcommittee of the federal Advisory Committee on Water Information (ACWI).

BOX 5-3
The Role of NAWQA's National Liaison
Committee in Cycle 3 Planning

In March 2010, the National Liaison Committee met to discuss planning for Cycle 3. The liaison committee was briefed on a preliminary draft of the Science Plan, which included the leadership vision for Cycle 3 and related science and policy questions the program planned to pursue. The committee was asked if this vision and related questions would meet the nation's needs in Cycle 3. Liaison committee members expressed strong support for:

- continued assessment of four major issues: excess nutrients, contaminants, sediment, and streamflow alteration;
- the planned rebuilding of the NAWQA status and trends networks in Cycle 3;
- coordinated water programs to leverage existing investments;
- a more robust national reference site network; and
- integration of monitoring, modeling, and understanding studies at multiple scales to forecast water-quality and ecosystem response to large-scale future changes (i.e., climate change and demographic change).

---

local efforts could be more consistent and perhaps beneficially enhanced. However, the design changes that have taken place during Cycle 2 have forced NAWQA away from the study unit framework to a more regional framework that does not generate the same level of local interest. As a result, there has been a corresponding decrease in study unit liaison committees. Focused topical studies, or understanding studies, have typically held liaison events, but attendance has been more limited compared to Cycle 1 because of the narrower scope of these studies (NAWQA leadership team, personal communication, May 2009). NAWQA's Major River Basins, Principal Aquifers, and Topical Study teams have also used stakeholder groups to review results and expected program reports and products.

The committee commends these ongoing efforts and encourages their continuance. **NAWQA should maintain its interface with the NLC and stakeholder groups, to the extent practical, to maintain these important relationships, thereby further leveraging resources to support collaborative efforts to implement the national Science Plan.**

The NRC's 2002 review of NAWQA, as it prepared for Cycle 2, offered other observations and recommendations on "Cooperation and Coordination Issues" that need not be restated here. Some of the recommendations became moot with design changes, as noted above. Yet NAWQA has made a significant, positive effort to address the key recommendations from NRC (2002), such as continued work on cooperative efforts with the USGS

National Research Program, the Toxics program, and the National Stream Quality Accounting Network (NASQAN).

COORDINATION AND COOPERATION
WITH EXTERNAL PARTNERS

NAWQA has worked to establish relationships with external partners beyond the NLC. The committee received testimony from many federal agency representatives and other stakeholders that highlighted interactions with and observations of NAWQA. From this dialogue and the committee's own observations, the committee concludes that NAWQA has done an admirable job of establishing collaborative relationships with other federal agencies and state-local authorities (Chapter 3). NAWQA's efforts have become critical to the missions of other agencies, and these relationships have strengthened NAWQA and USGS as a whole. NAWQA can use past experiences as models for the future efforts. Some examples follow, from the committee's assessment.

The U.S. Environmental Protection Agency

EPA is one of NAWQA's most critical partners. During Cycle 1 and part of Cycle 2, USGS placed staff as formal liaisons within several national offices of EPA to enhance coordination. USGS Water Resource Discipline staff liaisons were in residence in EPA's Office of Water and worked in support of the Safe Drinking Water Act (SDWA) Amendments of 1996,[5] the Clean Water Act,[6] and development of water-quality standards and criteria.[7] Other USGS staff were in residence in EPA's Office of Pesticide Programs (OPP), helping to provide information and technical support about pesticide occurrence in water, as well as fate and transport perspectives.

These USGS staff liaisons provided EPA with important scientific perspective in technical approaches to water-quality assessments, development of regional nutrient criteria, and identification of contaminants of emerging concern. Most importantly, perhaps, they also provided USGS with important perspectives on EPA's statutory responsibilities, for example, development of TMDLs and corresponding information needs. These staff liaisons enhanced working relations and led to some jointly funded projects that were complementary to NAWQA efforts (e.g., studies of pesticides in reservoirs used for drinking water). The formal liaisons were productive ac-

---

[5] The purview of EPA's Office of Ground Water and Drinking Water.
[6] The purview of the EPA's Office of Wetlands, Oceans, and Watersheds.
[7] The purview of EPA's Office of Science and Technology.

cording to testimony from agency personnel (NRC, 2009), but these liaison positions were terminated because of resource constraints.

Despite the termination of liaison positions, NAWQA has continued to coordinate with EPA. For example, NAWQA made substantive contributions to EPA's drinking water program in recent years. The 1996 amendments to the SDWA[8] called for EPA to develop new approaches to evaluating contaminants, old and new, that may need to be regulated in drinking water to protect public health. EPA's Office of Ground Water and Drinking Water is charged with developing a list of contaminants that may require regulation every 5 years, the Contaminant Candidate List (CCL), and developing a monitoring program, the Unregulated Contaminant Monitoring Regulation (UCMR). In addition, EPA must determine on a staggered 5-year deadline whether or not chemicals on the CCL warrant developing a regulation (the CCL-Regulatory Determination, or Reg Det process).

A key criterion for listing chemicals on the CCL and for the CCL-Reg Det process is whether the contaminant is known to occur or there is a substantial likelihood that the contaminant will occur in public water systems.[9] Data on actual occurrence of unregulated chemicals in finished drinking water are difficult to obtain, and EPA's authority is limited. The monitoring program (UCMR), for example, is limited in scope to no more than 30 contaminants every 5 years. Hence, NAWQA and USGS' Toxics program monitoring data have provided important insights on unregulated contaminants in ambient waters and in the source waters for drinking water systems that have been used in the CCL, CCL Reg-Det, and UCMR development processes. In turn, NAWQA has reviewed the CCL as it considered which contaminants to include in its own monitoring schedules. Of particular note, during the past 5 years, EPA implemented a more rigorous process to develop the third CCL (CCL 3) (EPA, 2009a, 2009b; NRC, 2001). NAWQA collaboratively provided EPA with Cycle 1 monitoring data so that EPA could evaluate the data to meet its specific CCL requirements. NAWQA staff have provided technical assistance to EPA programs as well as data for the Six-Year Review of regulated chemicals (EPA, 2009c). Despite this record of success, in testimony EPA representatives called for greater documentation and transparency in the selection of analytes to be monitored by NAWQA.

NAWQA data and cooperation have also contributed to the continuing efforts of EPA to meet the goals of the Clean Water Act. NAWQA occurrence data have been important to the prioritization of contaminants for development of water-quality criteria and aquatic life criteria by the Office

---

[8] Section 1445(a)(2); see 42 U.S.C. 300.
[9] See http://water.epa.gov/scitech/drinkingwater/dws/ccl/ccl3.cfm#overview.

of Science and Technology (OST) and the Office of Wetlands, Oceans, and Watersheds (OWOW). NAWQA has collaborated with these offices to promote standardized methods to states and other partners to help develop more uniform, comparable national data and to apply the SPARROW model to help identify areas on which to focus nutrient criteria and nutrient controls, as well as with states using these data for TMDL development (see also NRC, 2002).

NAWQA's national design provides a one-of-a-kind perspective not available in any other monitoring program.

Joseph Beaman, EPA OST, personal communication, September 21, 2009

NAWQA's data, as well as reviews by the U.S. Government Accountability Office and the National Research Council (NRC, 2002), have pointed out that EPA needed additional, different monitoring data to address the agency's performance. Neither NAWQA's design nor any other individual monitoring program can meet all needs. NAWQA staff consulted with and assisted EPA to develop its new monitoring approaches for the national Wadeable Stream Survey as part of the National Aquatic Resource Surveys. USGS has helped with planning discussions as well as consultations on site reconnaissance and sample collection, and supplemental data. NAWQA also supports EPA's *Report on the Environment* (EPA, 2008) to the Congress and the nation and international reviews of water issues (e.g., Global Water Research Coalition, 2004).

In addition, NAWQA staff and EPA Office of Research and Development (ORD) and OWOW have collaborated to combine EPA Environmental Monitoring and Assessment Program (EMAP)[10] data with NAWQA data to produce and publish aquatic models. In Cycle 1, NAWQA sampled ecological data at approximately 87 NAWQA sites, which were then augmented with a few hundred EMAP sites. Although the data had been collected with different field methods, researchers were able to develop a model to account for method bias, and the final ecological model was stronger for integrating both data sets (Carlisle and Hawkins 2008). EPA's ORD is putting resources into developing decision-support tools. Perhaps a specific effort between NAWQA's SPARROW effort and EPA would yield benefits for both. There has been some coordination between USGS and EPA on the initial planning for Mississippi River water-quality restoration with respect to nutrient loadings. Application of the USGS SPARROW

---

[10] See http://www.epa.gov/emap/.

model has been important in this regard, although opportunity for greater partnership between EPA and USGS exists (NRC, 2008b).

EPA's OPP has had a productive relationship with NAWQA for many years. The OPP uses NAWQA data and technical assistance to characterize the occurrence and trends of pesticides in water as part of risk assessments and implementation of the Food Quality Protection Act.[11] The OPP also uses NAWQA data directly in its Government Performance and Review Act, Program Assessment Rating Tool, which measures environmental outcomes of the OPP's programs. NAWQA has been of particular value because it is an independently derived national data set that can characterize pesticide trends and impacts on water quality (NRC, 2009).

The H. John Heinz III Center for Science, Economics, and the Environment

NAWQA provides collaborative technical support and data metrics for the *State of the Nation's Ecosystems 2008*, authored periodically by the H. John Heinz III Center for Science, Economics, and the Environment (H. John Heinz Center for Science, Economics, and the Environment, 2008). More than 20 of the Heinz Center ecological indicators were developed and are based on NAWQA data alone. At a public meeting of this committee, a Heinz Center representative speculated that the long-term integrity of USGS water-related data for as many as 25 national environmental indicators would be affected if USGS (NAWQA and the Geography Discipline) was unable to provide consistent data because of budget cuts (R. O'Malley, personal communication, September 21, 2009).

The Heinz Center depends heavily on NAWQA data to support our periodic report: *The State of the Nation's Ecosystems*. NAWQA data provide the foundation of our description of chemical contamination—including pesticides and other compounds—both nationally and among different land uses, and for tracking how contaminant levels change over time. We appreciate NAWQA's strong commitment to making its information and data readily accessible to meet our organization's needs and to address the Nation's water-resource information needs.

Robin O'Malley, Heinz Center Senior Fellow and Program Director, USGS Circular 1291

---

[11] The Food Quality Protection Act, passed in 1996, is pesticide food safety legislation.

Coordination and Cooperation with Other Agencies and Programs

NAWQA cooperates with many other agencies, including those in the public health arena. For example, the Centers for Disease Control and Prevention is collaborating with NAWQA to develop its National Environmental Public Health Tracking Network (EPHTN).[12] NAWQA provides water-quality information, particularly related to private drinking water supplies, to the state partners in the EPHTN (Bartholomay et al., 2007). A collaborative effort between NAWQA and the National Cancer Institute (NCI, part of the National Institutes of Health) developed an arsenic model for use in estimating exposure for NCI's New England Bladder Cancer Study (Ayotte et al., 2006a; Nuckols et al., 2011).[13] NAWQA and NCI also conducted an analysis of several locations exhibiting high incidences of cancer, using private-supply water use as a crude exposure term, region-by-region across the United States (Ayotte et al., 2006b). New England showed strong correlations to bladder, kidney, and lung cancer (Nuckols et al., 2011). In cooperation with the New Hampshire Environmental Public Health Tracking Network, part of EPHTN, the USGS New Hampshire—Vermont Water Science Center, with assistance from NAWQA, designed and developed a New Hampshire–specific arsenic model (publication forthcoming, J. Ayotte, personal communication, July 11, 2012).[14] This collaborative suite of efforts also involved researchers at the local and state levels, including representatives from Dartmouth's Geisel School of Medicine, the Departments of Health in New Hampshire and Vermont, and the Maine Center for Disease Control and Prevention.

NAWQA also provides cooperative support (e.g., technical expertise and data for water-quality indicators) to the ongoing development of the National Environmental Status and Trends Indicators project, a federal interagency project chaired by the U.S. Forest Service. The USGS Office of Water Quality, including NAWQA, provides funding and technical support for the National Water Quality Monitoring Council (NWQMC). This council consists of local, state, federal, privately funded, and volunteer organizations that "provide a forum to improve the nation's water quality through partnerships that foster increased understanding and stewardship of our water resources."[15] The NWQMC, with USGS and NAWQA input, have designed the National Monitoring Network for U.S. Coastal Waters

---

[12] See http://ephtracking.cdc.gov/showHome.action.

[13] NCI's epidemiological New England Bladder Cancer Study examines factors that might be associated with the high incidence of bladder cancer in the New England region.

[14] See http://www.nh.gov/epht/.

[15] See http://acwi.gov/monitoring/.

and Tributaries[16] that NAWQA supports through its surface water status and trends network.

NAWQA interacts with NOAA and EPA on estuaries and coastal water issues. NAWQA does not monitor and assess coastal waters and estuaries, in part because NOAA has responsibilities for estuaries and has assessment efforts and programs. EPA, in conjunction with NOAA, has established the National Estuary Program[17] that protects and restores estuaries of national significance. NAWQA collaborates with both of these agencies on coastal issues, particularly on matters related to growing concerns about eutrophication and hypoxia. NAWQA data provide the measures of nutrient and contaminant loading from upstream contributors into the estuaries. NAWQA has worked collaboratively with NOAA and EPA to apply and adapt models that assess the details of nutrient loading related to land use, management, and climate in major watersheds throughout the country. In particular, NAWQA has adapted and applied SPARROW to provide detailed information on the spatial distribution of sources in the Mississippi River basin that are delivering excess nutrients to the Gulf of Mexico. Based on this work, the Gulf of Mexico Hypoxia Task Force Action Plan was updated to look at regional targeting of management activities to work toward reducing nutrient loading.

The USDA Economic Research Service (ERS)[18] uses NAWQA products to evaluate the interactions between agriculture and water quality, particularly nutrients and pesticides. ERS relies on NAWQA synthesis reports to establish links between agriculture and observed regional water-quality and has adapted data and model coefficients from NAWQA and SPARROW to improve ERS models. These approaches are used by ERS to assess the economic efficiency, environmental effectiveness, and differential spatial and distributional implications of alternative agricultural policies and conservation practices that influence farm management decisions. In turn, ERS can then evaluate the impact of environmental policies and practices to protect water resources on the agricultural sector.

Despite the success of the aforementioned efforts, it has been difficult for NAWQA and USGS to establish significant relationships with NOAA, USDA, and other agencies like the U.S. Army Corps of Engineers (USACE) (NAWQA leadership, personal communication, September 21, 2009). In May 2011, during the committee's deliberations, NOAA, USACE, and USGS announced the signing of a Memorandum of Understanding (MOU) "to form an innovative partnership to address America's growing water re-

---

[16] See http://acwi.gov/monitoring/network/index.html.

[17] See http://water.epa.gov/type/oceb/nep/index.cfm.

[18] See http://www.ers.usda.gov/.

sources challenges."[19] This MOU appears to be a step toward more collaborative approaches, as urged for in this report. Similarly, USDA, EPA, and USGS are collectively working in coordination to implement and monitor projects under the USDA Natural Resources Conservation Service's "Mississippi River Basin Initiative" to improve water quality in the Mississippi River basin and reduce the impacts of the Gulf of Mexico hypoxia. Again, the committee hopes these efforts evolve into a productive cooperation.

The committee encourages USGS and NAWQA to use the new MOU with NOAA and USACE as an opportunity to define areas for cooperation to address key water-quality issues and to continue these and similar efforts, as defined in the Science Plan. For example, collaboration with researchers funded by the National Science Foundation's STReam Experimental and Observatory Network (STREON) program will enhance the level of understanding achievable when probing levels of nutrient enrichment that initiate ecological impairment (Objective 2c). Collaboration between EPA and NAWQA has yielded significant scientific value (see examples noted above). Maintaining regular contact with EPA's relevant program directors (e.g., Office of Water and Office of Pesticide Programs) and enhancing the interface with ORD's Safe and Sustainable Water Resources Research Program[20] would promote this relationship.

THE CHALLENGE AND IMPORTANCE
OF LOCAL RELATIONSHIPS

Coordination at the local level has been increasingly challenging given design alterations of the program (both planned and unplanned) as the role of study units has declined. To mitigate the loss of Study Unit Liaisons, NAWQA built stronger relationships with the USGS Water Science Centers during Cycle 2. The Water Science Centers also benefited from the development of projects that originate from efforts at the national perspective. For example, expanding upon the polycyclic aromatic hydrocarbons (PAH) national trends activity at the local level, the Water Science Center in Texas developed collaborative projects with the City of Austin, Texas. This effort documented significantly elevated PAH concentrations in residential areas and proposed that the source was coal tar–based sealcoat from parking lots. As a result, an understanding of the role of pavement sealcoat emerged, and the City of Austin banned the use of coal tar in sealcoat in 2006. Research continued to indicate that sealcoat is an important source of PAHs to the

---

[19] See http://www.noaanews.noaa.gov/stories2011/pdfs/usace_usgs_noaa_signmou.pdf, accessed March 2012.
[20] See http://www.epa.gov/aboutepa/ord/sswr.html.

environment, and a variety of actions followed to ban or restrict the use of sealcoat in the United States (Mahler et al., 2012).

The use of ancillary[21] data is becoming increasingly critical with the backdrop of dwindling federal resources. The use of ancillary data, when paired with modeling efforts, can extend NAWQA efforts into local areas without a NAWQA presence to bolster national coverage. To illustrate, the use of ancillary data has dramatically increased SPARROW coverage in the southeastern Major River Basin (MRB) (Figure 5-1). Furthermore, SPARROW's model error was reduced by 25 percent with the addition of these sites (NAWQA leadership, personal communication, October 26, 2010).

COLLABORATION IS ESSENTIAL IN CYCLE 3

In its current model of operations, NAWQA reaches out to other federal agencies, state and local governments, and the private sector (to a lesser degree) to seek their views of program priorities and useful products for decision-makers and the public. To further its program goals, NAWQA has developed some cooperative and collaborative relationships, coordinating data collection and analytical products with other organizations, and in some instances, with other USGS programs. Having noted this, the committee views NAWQA as functioning in Cycles 1 and 2 as primarily a self-contained federal program in which its own staff planned and conducted most elements of the monitoring and national syntheses. True collaboration (Box 5-1) takes this a step further, and is something the committee encourages the program to explore. However, the committee recognizes that collaboration as defined in Box 5-1 is probably more feasible within USGS than with other entities that have different data collection and analysis methods, congressional appropriations committees, and missions.

The Science Plan for Cycle 3 offers a comprehensive assessment of the nation's needs for understanding water-quality status and trends and for developing the models and analytical methods needed to understand and to forecast changes in water quality in response to changes in demography, land use, and climate. The Cycle 3 Science Plan presents a different vision and an expanded mission from the first two cycles and extends well beyond the capabilities and resources of NAWQA functioning in the largely autonomous mode it has historically used. **NAWQA should maintain its interface with the other federal agencies and stakeholder groups and work toward leveraging collaborative resources to meet the needs of the national Science Plan.** Quality assurance and quality control, with which NAWQA has

---

[21] Ancillary data are water-quality data collected by other USGS programs, national, regional, or local efforts on the same water-quality constituents monitored by NAWQA.

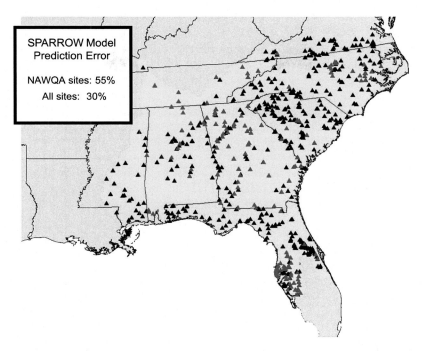

FIGURE 5-1 Integration of NAWQA data and ancillary data dramatically improves the spatial coverage of the SPARROW model. Red indicates USGS sites (196 sites), while black indicates state agency sites matched to a USGS gage (586 sites). Only 44 of the total sites shown are NAWQA monitoring sites. SOURCE: NAWQA leadership team, personal communication, May 9, 2009.

experience, will continue to be an issue in all cooperative and collaborative efforts; continued diligence is advised.

The Cycle 3 plan indirectly suggests that NAWQA will need to pay particular attention to its design in preparation for this more ambitious agenda. In the past, NAWQA has revamped its model operations as it has evolved—largely to cope with resource constraints—from the original study unit design to the Major River Basin "building blocks" to maintain a capacity to conduct national assessments. **To meet the national needs outlined in the Cycle 3 Science Plan, NAWQA will need to emphasize collaboration in two modes: as a leader that partners with other USGS and external programs, and as a follower with other federal agencies, state and local governments, and the private sector.**

As part of this approach, NAWQA will need to:

- focus on core mission areas where it has unique capabilities, for its own implementation efforts;
- leverage resources with other agencies to achieve more of the objectives of the Cycle 3 Science Plan;
- foster higher levels of involvement and investment by other agencies;
- help others design their own mission-critical programs to meet identified national objectives of the Cycle 3 Science Plan without NAWQA's direct involvement; and
- explore incentives, for example, access to NAWQA technical assistance, which will enable more sharing of effort for data collection, analysis, and technological innovation across the program.

The committee believes that advantages may exist in pursuing this approach. For example, NAWQA has a high concentration of water-quality analysts, and thus it may be able to offer technical assistance at lower cost than partners could procure through hiring or contracting on their own. At the same time, partners may have the capacity to collect field samples at less cost than NAWQA staff by virtue of their proximity to sampling sites and their flexibility to engage labor on an intermittent or part-time basis.

To operate in this more expansive mode, NAWQA should consider engaging partners and collaborators more directly in the development of mutual science plans, seamless exchanges of data and information, and joint implementation of work plans that identify shared responsibilities and accountability. Collaboration could be particularly critical at this point in time given the current fiscal climate and the continued decline of monitoring networks in the United States. Collaboration within USGS could serve as a starting point and a model between NAWQA and others outside USGS. NAWQA authored a forward-thinking comprehensive water-quality strategy for the nation during a climate of strained fiscal resources. This is an opportune time for NAWQA to bring together the federal agencies involved in water-quality monitoring and research and, using the Science Plan as a starting point, to develop a collaborative water-quality strategy for the nation.

References

Alexander, R. B., R. A. Smith, and G. E. Schwarz. 2000. Effect of stream channel size on the delivery of nitrogen to the Gulf of Mexico. Nature 403:758-761.

Alexander, R. B., R. A. Smith, G. E. Schwartz, E. W. Boyer, J. V. Nolan, and J. W. Brakebill. 2008. Differences in phosphorus and nitrogen delivery to the Gulf of Mexico from the Mississippi River basin. Environmental Science and Technology 42:822-830.

Anning, D. W., N. J. Bauch, S. J. Gerner, M. E. Flynn, S. N. Hamlin, S. J. Moore, D. H. Schaefer, S. K. Anderholm, and L. E. Spangler. 2007. Dissolved Solids in Basin-Fill Aquifers and Streams in the Southwestern United States. U.S. Geological Survey Scientific Investigations Report 2006-5315.

Ayotte, J. D., B. T. Nolan, J. R. Nuckols, K. P. Cantor, G. R. Robinson, D. Baris, L. Hayes, M. Karagas, W. Bress, D. T. Silverman, and J. H. Lubin. 2006a. Modeling the probability of arsenic in groundwater in New England as a tool for exposure assessment. Environmental Science and Technology 40(11):3578-3585.

Ayotte, J. D., D. Baris, K. P. Cantor, J. Colt, G. R. Robinson Jr., J. H. Lubin, M. Karagas, R. N. Hoover, J. F. Fraumeni Jr., and D. T. Silverman. 2006b. Bladder cancer mortality and private well use in New England: An ecological study. Journal of Epidemiology & Community Health 60(2):168-172.

Bartholomay, R. C., J. M. Carter, S. L. Qi, P. J. Squillace, and G. L. Rowe. 2007. Summary of Selected U.S. Geological Survey Data on Domestic Well Water Quality for the Centers for Disease Control's National Environmental Public Health Tracking Program. U.S. Geological Survey Scientific Investigations Report 2007-5213. 57 pp.

Bauch, N. J., L. C. Chasar, B. C. Scudder, P. W. Moran, K. J. Hitt, M. E. Brigham, M. A. Lutz, and D. A. Wentz. 2009. Data on Mercury in Water, Streambed Sediment, and Fish Tissue from Selected Streams Across the United States, 1998–2005. U.S. Geological Survey Data Series Report 307.

Bell, R. W., and A. K. Williamson. 2006. Data Delivery and Mapping Over the Web: National Water-Quality Assessment Data Warehouse. USGS Fact Sheet 2006-3101.

Booth, D. B., and C. J. Jackson. 1997. Urbanization of aquatic systems: Degradation thresholds, stormwater detention and the limits of mitigations. Water Resources Bulletin 33:1077-1090.

Booth, N. L., E. J. Everman, I. Lin Kuo, L. Sprague, and L. Murphy. 2011. A web-based decision support system for assessing regional water-quality conditions and management actions. Journal of the American Water Resources Association 47(5):1136-1150.

Brakebill, J. W., S. W. Ator, and G. E. Schwarz. 2010. Sources of suspended-sediment flux in streams of the Chesapeake Bay watershed: A regional application of the SPARROW model. Journal of the American Water Resources Association 46:757-776, doi: 10.1111/j.1752-1688.2010.00450.x.

Brezonik, P. L., V. J. Bierman, R. Alexander, J. Anderson, J. Barko, M. Dortch, L. Hatch, D. Keeney, D. Mulla, V. Smith, C. Walker, T. Whitledge, and W. Wiseman. 1999. Effects of reducing nutrient loads to surface waters within the Mississippi River basin and the Gulf of Mexico. Report of Task Group 4 to the White House Committee on Environment and Natural Resources, Hypoxia Work Group. Federal Register 64:23834-23835. Available online at http://oceanservice.noaa.gov/products/pubs_hypox.html.

Brigham, M. E., D. A. Wentz, G. R. Aiken, and D. P. Krabbenhoft. 2009. Mercury cycling in stream ecosystems. 1. Water column chemistry and transport. Environmental Science and Technology 43(8):2720-2725.

Brown, L. R., T. F. Cuffney, J. F. Coles, F. Fitzpatrick, G. McMahon, J. Steuer, A. H. Bell, and J. T. May. 2009. Urban streams across the USA: Lessons learned from studies in 9 metropolitan areas. Journal of the North American Benthological Society 28:1051-1069.

Carlisle, D. M., and C. P. Hawkins. 2008. Land use and the structure of western US stream invertebrate assemblages: Predictive models and ecological traits. Journal of the North American Benthological Society 27:986-999.

Carlisle, D. M., J. Falcone, D. M. Wolock, M. R. Meador, and R. H. Norris. 2009. Predicting the natural flow regime: Models for assessing hydrological alteration in streams. River Research and Applications 26:118-136, doi: 10.1002/rra.1247.

Carlisle, D. M., D. M. Wolock, and M. R. Meador. 2011. Alteration of streamflow magnitudes and potential ecological consequences: A multiregional assessment. Frontiers in Ecology and the Environment 9:264-270.

Carter, J. M., G. C. Delzer, J. A. Kingsbury, and J. A. Hopple. 2007. Concentration Data for Anthropogenic Organic Compounds in Ground Water, Surface Water, and Finished Water of Selected Community Water Systems in the United States, 2002–05: U.S. Geological Survey Data Series 268. 30 pp.

Chalmers, A. T., D. M. Argue, D. A. Gay, M. E. Brigham, C. J. Schmitt, and D. L. Lorenz. 2010. Mercury trends in fish from rivers and lakes in the United States, 1969–2005. Environmental Monitoring and Assessment 175(1-4):175-191, doi: 10.1007/s10661-010-1504-6.

Chasar, L. C., B. C. Scudder, A. R. Stewart, A. H. Bell, and G. R. Aiken. 2009. Mercury cycling in stream ecosystems. 3. Trophic dynamics and methylmercury bioaccumulation. Environmental Science and Technology 43(8):2733-2739.

Crawford, C., P. Hamilton, and A. Hoos. 2006. National Water-Quality Assessment Program—Modifications to the Status and Trends Network and Assessments of Streams and Rivers. Prepared by Charlie Crawford, Pixie Hamilton, and Anne Hoos for a NAWQA National Liaison Meeting, Washington D.C., October 5, 2006. Available online at http://water.usgs.gov/nawqa/studies/mrb/mrb_factsheet.pdf.

Cuffney, T. E., H. Zappia, E. M. P. Giddings and J. F. Coles. 2005. Effects of urbanization on benthic macroinvertebrate assemblages in contrasting environmental settings: Boston, Massachusetts; Birmingham, Alabama; and Salt Lake City, Utah. American Fisheries Society Symposium 47:361-407.

Cuffney, T. F., and J. A. Falcone. 2009. Derivation of Nationally Consistent Indices Representing Urban Intensity Within and Across Nine Metropolitan Areas of the Conterminous United States. U.S. Geological Survey Scientific Investigations Report 2008-5095. 36 pp.

Delzer, G. C., and P. A. Hamilton. 2007. National Water-Quality Assessment Program—Source Water-Quality Assessments. U.S. Geological Survey Fact Sheet 2007-3069. 2 pp.

Diaz, R. J., and R. Rosenberg. 2008. Spreading dead zones and consequences for marine ecosystems. Science 32:926-929.

Dubrovsky, N. M., K. R. Burow, G. M. Clark, J. M. Gronberg, P. A. Hamilton, K. J. Hitt, D. K. Mueller, M. D. Munn, B. T. Nolan, L. J. Puckett, M. G. Rupert, T. M. Short, N. E. Spahr, L. A. Sprague, and W. G. Wilber. 2010. The quality of our Nation's waters—Nutrients in the Nation's streams and groundwater, 1992–2004: U.S. Geological Survey Circular 1350, 174 pp.

Duff, J. H., A. J. Tesoriero, W. B. Richardson, E. A. Strauss, and M. D. Munn. 2008. Whole-stream response to nitrate loading in three streams draining agricultural landscapes. Journal of Environmental Quality 37(3):1133-1144.

Ehberts, S. M., J. K. Böhlke, L. J. Kauffman, and B. C. Jurgens. 2012. Comparison of particle-tracking and lumped-parameter age-distribution models for evaluating vulnerability of production wells to contamination. Hydrogeology Journal 20:263-282.

Embrey, S. S., and D. L. Runkle. 2006. Microbial Quality of the Nation's Ground-Water Resources, 1993-2004. U.S. Geological Survey Scientific Investigations Report 2006-5290. 34 pp.

Entekhabi, D., G. R. Asrar, A. K. Betts, K. J. Beven, R. L. Bras, C. J. Duffy, T. Dunne, R. D. Koster, D. P. Lettenmaier, D. B. McLaughlin, W. J. Shuttleworth, M. T. van Genuchten, M. Y. Wei, and E. F. Wood. 1999. An agenda for land surface hydrology research and a call for the second international hydrological decade. Bulletin of the American Meteorological Society 80:2043-2058.

EPA (U.S. Environmental Protection Agency). 2008. EPA's 2008 Report on the Environment. National Center for Environmental Assessment, Washington, DC. EPA/600/R-07/045F. Available from the National Technical Information Service, Springfield, VA, and online at http://www.epa.gov/roe.

EPA. 2009a. Drinking water contaminant candidate list 3–final. Federal Register 74(194):51850.

EPA. 2009b. Final Contaminant Candidate List 3 Chemicals: Classification of PCCL to the CCL. EPA 815-R-09-008. August.

EPA. 2009c. Analysis of Occurrence Data from the Second Six-Year Review of Existing National Primary Drinking Water Regulations. U.S. EPA Report 815-B-09-006. Available online at http://water.epa.gov/lawsregs/rulesregs/regulatingcontaminants/sixyearreview/second_review/index.cfm.

Essaid, H. I., C. M. Zamora, K. A. McCarthy, J. R. Vogel, and J. T. Wilson. 2008. Using heat to characterize streambed water flux variability in four stream reaches. Journal of Environmental Quality 37:1010-1023.

Francy, D. S., D. H. Helsel, and R. A. Nally. 2000. Occurrence and distribution of microbiological indicators in ground water and stream water. Water Environment Research 72(2):152-161.

Frankforter, J. D., H. S. Weyers, J. D. Bales, P. W. Moran, and D. L. Calhoun. 2009. The relative influence of nutrients and habitat on stream metabolism in agricultural streams. Environmental Management and Assessment 168(1-4):461-479, doi: 10.1007/s10661-009-1127-y.

Gibert, J., M. J. Dole-Olivier, P. Marmonier, and P. Vervier. 1990. Surface water/groundwater ecotones. Pp. 199-225 in Ecology and Management of Aquatic-Terrestrial Ecotones, Man and the Biosphere Series, Vol. 4, edited by R. J. Naiman and H. Decamps. Paris: United Nations Educational, Scientific, and Cultural Organization.

Giddings, E. M. P., A. H. Bell, K. M. Beaulieu, T. F. Cuffney, J. F. Coles, L. R. Brown, F. A. Fitzpatrick, J. Falcone, L. A. Sprague, W. L. Bryant, M. C. Peppler, C. Stephens, and G. McMahon. 2009. Selected Physical, Chemical, and Biological Data Used to Study Urbanizing Streams in Nine Metropolitan Areas of the United States, 1999–2004. U.S. Geological Survey Data Series 423. 11 pp.

Gilliom, R. J., W. M. Alley, and M. E. Gurtz. 1995. Design of the National Water-Quality Assessment Program: Occurrence and Distribution of Water-Quality Conditions. U.S. Geological Survey Circular 1112. 33 pp.

Gilliom, R. J., K. Bencala, C. A. Couch, D. Helsel, W. W. Lapham, J. Stoner, W. G. Wilbur, J. Zogorski, W. Bryant, N. M. Dubrovsky, L. Franke, I. James, D. Mueller, M. A. Sylvester, and D. M. Wolock. 2000. Study-Unit Design Guidelines for Cycle II of the National Water Quality Assessment (NAWQA). U.S. Geological Survey NAWQA Cycle II Implementation Team. Draft for internal review (June 28, 2001). Sacramento, CA: U.S. Geological Survey.

Gilliom, R. J., P. A. Hamilton, and T. L. Miller. 2001. The National Water Quality Assessment Program—Entering a New Decade of Investigations. U.S. Geological Survey Fact Sheet 071-01. Reston, VA: U.S. Geological Survey. Available online at http://pubs.usgs.gov/fs/fs-071-01/pdf/fs07101.pdf.

Gilliom, R. J., J. E. Barbash, C. G. Crawford, P. A. Hamilton, J. D. Martin, N. Nakagaki, L. H. Nowell, J. C. Scott, P. E. Stackelberg, G. P. Thelin, and D. M. Wolock. 2006. Pesticides in the Nation's Streams and Ground Water, 1992-2001. U.S. Geological Survey Circular 1291.

Global Water Research Coalition. 2004. Pharmaceuticals and Personal Care Products in the Water Cycle: Report of the Global Water Research Coalition Research Strategy Workshop. International Water Association. Kiwa Water Research and Stowa (Netherlands). 72 pp.

Gurdak, J. J., P. B. McMahon, K. F. Dennehy, and S. L. Qi. 2009. Water Quality in the High Plains Aquifer, Colorado, Kansas, Nebraska, New Mexico, Oklahoma, South Dakota, Texas, and Wyoming, 1999–2004. U.S. Geological Survey Circular 1337. 63 pp.

H. John Heinz Center for Science, Economics, and the Environment. 2008. The State of the Nation's Ecosystems. Measuring the Lands, Waters, and Living Resources of the United States. Washington, DC: Island Press.

Hamilton, P. A., and R. J. Shedlock. 1992. Are Fertilizers and Pesticides in the Ground Water—A Case Study of the Delmarva Peninsula, Delaware, Maryland, and Virginia: U.S. Geological Survey Circular 1080. 16 pp.

Hamilton, P. A., T. L. Miller, and D. N. Myers. 2004. Water Quality in the Nation's Streams and Aquifers—Overview of Selected Findings, 1991–2001. U.S. Geological Survey Circular 1265.

Harding, L. W., Jr., D. Degobbis, and R. Precali. 1999. Production and fate of phytoplankton: Annual cycles and interannual variability. Pp. 131-172 in Coastal and Estuarine Studies: Ecosystems at the Land-Sea Margin Drainage Basin to Coastal Sea, edited by T. C. Malone, A. Malej, L. W. Harding Jr., N. Smodlaka, and R. E. Turner. Washington, D.C.: American Geophysical Union.

Helly, J. J., and L. A. Levin. 2004. Global distribution of naturally occurring marine hypoxia on continental margins. Deep Sea Research Part I. Oceanographic Research Papers 51:1159-1168.

Hoos, A. N., and G. McMahon. 2009. Spatial analysis of instream nitrogen loads and factors controlling nitrogen delivery to streams in the southeastern United States using spatially reference regression on watershed attributes (SPARROW) and regional classification frameworks. Hydrological Processes 23(16):2275-2294, doi: 10.1002/hyp.7323.

Hopple, J. A., G. C. Delzer, and J. A. Kingsbury. 2009. Anthropogenic organic compounds in source water of selected community water systems that use groundwater, 2002–05. U.S. Geological Survey Scientific Investigations Report 2009-5200. 74 pp.

Justus, B. G., J. C. Petersen, S. R. Femmer, J. V. Davis, and J. E. Wallace. 2010. A comparison of algal, macroinvertebrate, and fish assemblage indices for assessing low-level nutrient enrichment in wadeable Ozark streams. Ecological Indicators 10(3):627-638.

Kashuba, R., Y. K. Cha, I. Alameddine, B. Lee, and T. F. Cuffney. 2010. Multilevel Hierarchical Modeling of Benthic Macroinvertebrate Responses to Urbanization in Nine Metropolitan Regions Across the Conterminous United States. U.S. Geological Survey Scientific Investigations Report 2009–5243. 88 pp.

Kaushal, S. S., P. M. Groffman, G. E. Likens, K. T. Belt, W. P. Stack, V. R. Kelly, L. E. Band, and G. T. Fisher. 2005. Increased salinization of freshwater in the northeastern United States. Proceedings of the National Academy of Sciences of the United States of America 102:13517-13520.

Kingsbury, J. A., G. C. Delzer, and J. A. Hopple. 2008. Anthropogenic organic compounds in source water of nine community water systems that withdraw from streams, 2002–05. U.S. Geological Survey Scientific Investigations Report 2008-5208. 66 pp.

Knopman, D. S., and R. A. Smith. 1993. Twenty years of the Clean Water Act: Has U.S. water quality improved? Environment 35(1):17-34.

Kolpin, D. W., E. T. Furlong, M. T. Meyer, E. M. Thurman, S. D. Zaugg, L. B. Barber, and H. T. Buxton. 2002. Pharmaceuticals, hormones, and other organic wastewater contaminants in U.S. streams, 1999-2000—A national reconnaissance. Environmental Science and Technology 36(6):1202-1211.

Lapham, W., P. Hamilton, and D. Myers. 2005. National Water-Quality Assessment Program—Cycle II. Regional Assessment of Aquifers. U.S. Geological Survey Fact Sheet 071-01. Reston, VA: U.S. Geological Survey. Available online at http://pubs.usgs.gov/fs/2005/3013/pdf/PASforWeb.pdf.

Larson, S. J., and R. J. Gilliom. 2001. Regression models for estimating herbicide concentrations in U.S. streams from watershed characteristics. Journal of the American Water Resources Association 37(5):1349-1368.

Larson, S. J., C. G. Crawford, and R. J. Gilliom. 2004. Development and Application of Watershed Regressions for Pesticides (WARP) for Estimating Atrazine Concentration Distributions in Streams. U.S. Geological Survey Water-Resources Investigations Report 03-4047. 68 pp.

Lorenz, D. L., D. M. Robertson, and D. W. Hall. 2009. Trends in Streamflow, and Nutrient and Suspended Sediment Concentrations and Loads in the Upper Mississippi, Ohio, Red, and Great Lakes River Basins 1975-2004. U.S. Geological Survey Scientific Investigations Report 2008-5213. 82 pp.

Mahler, B. J., P. C. Van Metre, and E. Callender. 2006. Trends in metals in urban and reference lake sediments across the United States, 1970-2001. Environmental Toxicology and Chemistry 25(7):1698-1709.

Mahler, B. J., P. C. Van Metre, J. L. Crane, A. W. Watts, M. Scoggins, and E. S. Williams. 2012. Coal-tar-based pavement sealcoat and PAHs: Implications for the environment, human health, and stormwater management. Environmental Science and Technology 46:3039-3045.

Maret, T. R., C. P. Konrad, and A. W. Tranmer. 2010. Influence of environmental factors on biotic responses to nutrient enrichment in agricultural streams. Journal of the American Water Resources Association 46(3):498-513, doi: 10.1111/j.1752-1688.2010.00430.x.

Marvin-DiPasquale, M., M. A. Lutz, M. E. Brigham, D. P. Krabbenhoft, G. R. Aiken, W. H. Orem, and B. D. Hall. 2009. Mercury cycling in stream ecosystems. 2. Benthic methylmercury production and bed sediment—pore water partitioning. Environmental Science and Technology 43(8):2726-2732.

McCarthy, K. A. 2009. A Whole-System Approach to Understanding Agricultural Chemicals in the Environment. U.S. Geological Survey Fact Sheet 2009-3042. 6 pp. Available online at http://pubs.usgs.gov/fs/2009/3042/.

McMahon, G., and T. F. Cuffney. 2000. Quantifying urban intensity in drainage basins for assessing stream ecological conditions. Journal of the American Water Resources Association 36:1247-1261.

McMahon, P. B., K. F. Dennehy, B. W. Bruce, J. J. Gurdak, and S. L. Qi. 2007. Water-Quality Assessment of the High Plains Aquifer, 1999–2004. U.S. Geological Survey Professional Paper 1749. 136 pp.

McMahon, P. B., K. R. Burow, L. J. Kauffman, S. M. Eberts, J. K. Bohlke, and J. J. Gurdak. 2008. Simulated response of water quality in public supply wells to land use change, Water Resources Research 44:W00A06, doi: 10.1029/2007WR006731.

Milly, P. C. D., J. Betancourt, M. Falkenmark, R. M. Hirsch, Z. W. Kundzewicz, D. P. Lettenmaier, and R. J. Stouffer. 2008. Stationarity is dead: Whither water management? Science 319:573-574.

Moorman, M. C., D. A. Harned, G. McMahon, and K. Capelli. 2011. Improving scientific communication through the use of U.S. Geological Survey video podcasts. Proceedings of the 2011 George Wright Society Conference on Parks, Protected Areas, and Cultural Sites. Available online at http://www.georgewright.org/1141moorman.pdf.

Moran, M. J., W. W. Lapham, B. L. Rowe, and J. S. Zogorski. 2004. Volatile organic compounds in ground water from rural private wells, 1986-1999. Journal of the American Water Resources Association 40(5):1141-1157.

Moran, M. J., J. S. Zogorski, and P. J. Squillace. 2005. MTBE and gasoline hydrocarbons in ground water of the United States. Ground Water 43(4):615-627.

Moran, M. J., J. S. Zogorski, and P. J. Squillace. 2007. Chlorinated solvents in groundwater of the United States. Environmental Science and Technology 41(1):74-81.

Munn, M. D., and P. A. Hamilton. 2003. New Studies Initiated by the U.S. Geological Survey—Effects of Nutrient Enrichment on Stream Ecosystems. U.S. Geological Survey Fact Sheet 118-03. 4 pp.

National Science and Technology Council, Committee on Environment and Natural Resources, Subcommittee on Water Availability and Quality. 2007. A Strategy for Federal Science and Technology to Support Water Availability and Quality in the United States. 46 pp.

Nolan, B. T., and K. J. Hitt. 2006. Vulnerability of shallow groundwater and drinking-water wells to nitrate in the United States. Environmental Science and Technology 40:7834-7840.

NRC (National Research Council). 1985. Letter Report on a Proposed National Water Quality Assessment Program. Washington, DC: National Academy Press.

NRC. 1987. National Water Quality Monitoring and Assessment. Report on a Colloquium Sponsored by the Water Science and Technology Board, May 21-22, 1986. Washington, DC: National Academy Press.

NRC. 1990. A Review of the USGS National Water Quality Assessment Pilot Program. Washington, DC: National Academy Press.

NRC. 1991. Opportunities in the Hydrologic Sciences. Washington, DC: National Academy Press.

NRC. 1994. National Water Quality Assessment Program: The Challenge of National Synthesis. Washington, DC: National Academy Press.

NRC. 2001. Classifying Drinking Water Contaminants for Regulatory Considerations. Washington, DC: National Academy Press.

NRC. 2002. Opportunities to Improve the U.S. Geological Survey National Water Quality Assessment Program. Washington, DC: National Academies Press.

NRC. 2004a. Nonnative Oysters in the Chesapeake Bay. Washington, DC: National Academies Press.

NRC. 2004b. Confronting the Nation's Water Problems: The Role of Research. Washington, DC: National Academies Press.

NRC. 2008a. Progress Toward Restoring the Everglades: The Second Biennial Review. Washington, DC: National Academies Press.

NRC. 2008b. Mississippi River Water Quality and the Clean Water Act: Progress, Challenges, and Opportunities. Washington, DC: National Academies Press.

NRC. 2009. Toward a Sustainable and Secure Water Future: A Leadership Role for the U.S. Geological Survey. Washington, DC: National Academies Press.

NRC. 2010. Letter Report Assessing the USGS National Water Quality Assessment Program's Science Framework. Washington, DC: National Academies Press.

NRC. 2011a. Letter Report Assessing the USGS National Water Quality Assessment Program's Science Plan. Washington, DC: National Academies Press.

NRC. 2011b. Achieving Nutrient and Sediment Reduction Goals in the Chesapeake Bay: An Evaluation of Program Strategies and Implementation. Washington, DC: National Academies Press.

NRC. 2011c. A Review of the Use of Science and Adaptive Management in California's Draft Bay Delta Conservation Plan. Washington, DC: National Academies Press.

NRC. 2012. Challenges and Opportunities in the Hydrologic Sciences. Washington, DC: National Academies Press.

Nuckols, J. R., L. E. Beane Freeman, J. H. Lubin, M. S. Airola, D. Baris, J. D. Ayotte, A. Taylor, C. Paulu, M. R. Karagas, J. Colt, M. H. Ward, A.-T. Huang, W. Bress, S. Cherala, D. T. Silverman, and K. P. Cantor. 2011. Estimating water supply arsenic levels in the New England bladder cancer study. Environmental Health Perspectives 119(9):1279-1285.

Osterman, L. E., P. W. Swarzenski, and R. Z. Poore. 2006. Gulf of Mexico Dead Zone—The Last 150 Years. U.S. Geological Survey Fact Sheet 2006-3005.

Pasquale, M. M. D., M. A. Lutz, M. E. Brigham, D. P. Krabbenhoft, G. R. Aiken, W. H. Orem, and B. D. Hall. 2009. Mercury cycling in stream ecosystems. 2. Benthic methylmercury production and bed sediment-pore water partitioning. Environmental Science and Technology 43(8):2726-2732.

Phillips, P. J., S. W. Ator, and E. A. Nystrom. 2007. Temporal changes in surface-water insecticide concentrations after the phaseout of diazinon and chlorpyrifos. Environmental Science and Technology 41(12):4246-4251, doi: 10.1021/es070301.

Poff, N. L., J. D. Allan, M. B. Bain, J. R. Karr, K. L. Prestegaard, B. D. Richter, R. E. Sparks, and J. C. Stromberg. 1997. The natural flow regime: A paradigm for river conservation and restoration. BioScience 47:769-784.

Puckett, L. J., C. Zamora, H. Essaid, J. T. Wilson, H. M. Johnson, M. J. Brayton and J. R. Vogel. 2008. Transport and fate of nitrate at the ground-water/surface-water interface. Journal of Environmental Quality 37:1034-1050.

Robertson, D. M., G. E. Schwarz, D. A. Saad, and R. B. Alexander. 2009. Incorporating uncertainty into the ranking of SPARROW model nutrient yields from the Mississippi/Atchafalaya River Basin watersheds. Journal of the American Water Resources Association 45(2):534-549, doi: 10.1111/j.1752-1688.2009.00310.x.

Ryker, S. J., and A. K. Williamson. 1996. Pesticides in Public Supply Wells of Washington State. U.S. Geological Survey Fact Sheet 122-96.

Schoups, G., J. W. Hopmans, C. A.Young, J. A. Vrugt, W. W. Wallender, J. K. Tanji, and S. Panday. 2005. Sustainability of irrigated agriculture in the San Joaquin Valley, California. Proceedings of the National Academy of Sciences 102:15352-15356.

Schueler, T. R. 1994. The importance of imperviousness. Watershed Protection Techniques 1:73-75.

Scudder, B. C., L. C. Chasar, D. A. Wentz, N. J. Bauch, M. E. Brigham, P. W. Moran, and D. P. Krabbenhoft. 2009. Mercury in fish, bed sediment, and water from streams across the United States, 1998–2005. U.S. Geological Survey Scientific Investigations Report 2009-5109. 74 pp.

Smith, R. A., G. E. Schwarz, and R. B. Alexander. 1997. Regional interpretation of water-quality monitoring data. Water Resources Research 33(12):2781-2798.

Smith, R. A., R. B. Alexander, G. E. Schwartz. 2003. Natural background concentrations of nutrients in streams and rivers of the conterminous United States. Environmental Science and Technology 37(14):3039-3047.

Sprague, L. A., M. L. Clark, D. L. Rus, R. B. Zelt, J. L. Flynn, and J. V. Davis. 2007. Nutrient and Suspended-Sediment Trends in the Missouri River Basin, 1993-2003. U.S. Geological Survey Scientific Investigations Report 2006-5231. 80 pp.

Stone, W. W., and R. J. Gilliom. 2009. Update of Watershed Regressions for Pesticides (WARP) for Predicting Atrazine Concentration in Streams. U.S. Geological Survey Open-File Report 2009-1122. 22 pp.

Stone, W. W., and R. J. Gilliom. 2012. Watershed regressions for pesticides (WARP) models for predicting atrazine concentrations in corn belt streams. Journal of the American Water Resources Association 1-17, doi: 10.1111/j.1752-1688.2012.00661.x.

Stone, W. W., R. J. Gilliom, and C. G. Crawford. 2008. Watershed Regressions for Pesticides (WARP) for Predicting Annual Maximum and Maximum Moving-Average Concentrations of Atrazine in Streams. U.S. Geological Survey Open-File Report 08-1186. 19 pp.

Sullivan, D. J., A. V. Vecchia, D. L. Lorenz, R. J. Gilliom, and J. D. Martin. 2009. Trends in Pesticide Concentration in Corn-Belt Streams, 1996-2006. U.S. Geological Survey Scientific Investigations Report 2009-5132. 75 pp.

Tate, C. M., T. F. Cuffney, G. McMahon, E. M. P. Giddings, J. E. Coles, and H. Zappia. 2005. Use of an urban intensity index to assess urban effects on streams in three contrasting environmental settings. American Fisheries Society Symposium 47:291-315.

Tesoriero, A. J., D. A. Saad, K. R. Burow, E. A. Frick, L. J. Puckett, and J. E. Barbash. 2007. Linking ground-water age and chemistry data along flow paths: Implications for trends and transformations of nitrate and pesticides. Journal of Contaminant Hydrology 94(1-2):139-155.

USGS (U.S. Geological Survey). variously dated. National field manual for the collection of water-quality data: U.S. Geological Survey Techniques of Water-Resources Investigations, book 9, chapters A1-A9. Available online at http://pubs.water.usgs.gov/twri9A.

USGS. 1984. The National Water Summary 1983—Hydrologic Events and Issues. U.S. Geological Survey Water Supply Paper 2250.

USGS. 1999. The Quality of Our Nation's Waters—Nutrients and Pesticides. Circular 1225. Reston, VA: USGS. 82 pp.

USGS. 2001. The National Water Quality Assessment Program—Informing Water-Resource Management and Protection Decision. Available online at http://water.usgs.gov/nawqa/docs/xrel/external.relevance.pdf.

USGS. 2005. Sustainability National Water-Quality Assessment Program—Cycle II: Regional Assessments of Aquifers. Fact Sheet 2005-3013. Available online at http://pubs.usgs.gov/fs/2005/3013/.

USGS. 2007. Facing Tomorrow's Challenges—U.S. Geological Survey Science in the Decade, 2007-2017. U.S. Geological Survey Circular 1309. 69 pp. Available online at http://pubs.usgs.gov/circ/2007/1309/.

USGS. 2008. The National Water-Quality Assessment Program—Progress To-Date and Setting the Stage for the Future. Briefing Sheet. Available online at http://water.usgs.gov/nawqa/BriefingSheet.20081009.pdf.

USGS. 2009a. Mercury in Aquatic Ecosystems—Recent Findings from the National Water-Quality Assessment (NAWQA) and Toxic Substances Hydrology Programs (as presented to the National Liaison Committee, August 21, 2009). Available online at http://water.usgs.gov/nawqa/mercury/mercury.handout.final.08172009.pdf.

USGS. 2009b. SPARROW MODELING—Enhancing Understanding of the Nation's Water Quality. Fact Sheet 2009-3019. Available online at http://pubs.usgs.gov/fs/2009/3019/.

USGS. 2010. The National Water-Quality Assessment Program—Science to Policy and Management. Available online at http://water.usgs.gov/nawqa/xrel.pdf.

Van Metre, P. C., and B. J. Mahler. 2005. Trends in hydrophobic organic contaminants in urban and reference lake sediments across the United States, 1970-2001. Environmental Science and Technology 39(15):5567-5574.

Van Metre, P. C., B. J. Mahler, and J. T. Wilson. 2009. PAHs underfoot: Contaminated dust from coal-tar sealcoated pavement is widespread in the United States. Environmental Science and Technology 43(1):20-25, doi: 10.1021/es802119h.

Vervier, P., J. Gibert, P. Marmonier, and M. J. Dole-Olivier. 1992. A perspective on the permeability of the surface freshwater-groundwater ecotone. Journal of the North American Benthological Society 11:93-102.

Vowinkel, E. F., R. M. Clawges, D. E. Buxton, D. A. Stedfast, and J. B. Louis. 1996. Vulnerability of Public Drinking Water Supplies in New Jersey to Pesticides. U.S. Geological Survey Fact Sheet 165-96.

Wenger, S., A. H. Roy, C. R. Jackson, E. S. Bernhardt, T. L. Carter, S. Filoso, C. A. Gibson, W. C. Hession, S. S. Kaushal, E. Marti, J. L. Meyer, M. A. Palmer, M. J. Paul, A. H. Purcell, A. Ramírez, A. D. Rosemond, K. A. Schofield, E. B. Sudduth, and C. J. Walsh. 2009. Twenty-six key research questions in urban stream ecology: An assessment of the state of the science. Journal of the North American Benthological Society 28:1080-1098.

Winter, T. C. 2001. The concept of hydrologic landscapes. Journal of the American Water Resources Association 37(2):335-349.

Wise, D. R., F. A. Rinella III, J. F. Rinella, G. J. Fuhrer, S. S. Embrey, G. E. Clark, G. E. Schwarz, and S. Sobieszczyk. 2007. Nutrient and Suspended-Sediment Transport and Trends in the Columbia River and Puget Sound Basins, 1993-2003. U.S. Geological Survey Scientific Investigations Report 2007-5186. 177 pp.

Zogorski, J. S., J. M. Carter, T. Ivahnenko, W. W. Lapham, M. J. Moran, B. L. Rowe, P. J. Squillace, and P. L. Toccalino. 2006. The Quality of our Nation's Waters—Volatile Organic Compounds in the Nation's Ground Water and Drinking-Water Supply Wells. U.S. Geological Survey Circular 1292. 101 pp.

Appendix A

Letter Report Assessing the USGS National Water Quality Assessment Program's Science Framework

Letter Report Assessing the USGS National Water Quality Assessment Program's Science Framework

Committee on Preparing for the Third Decade (Cycle 3)
of the National Water Quality Assessment (NAWQA) Program

Water Science and Technology Board

Division on Earth and Life Studies

NATIONAL RESEARCH COUNCIL
OF THE NATIONAL ACADEMIES

THE NATIONAL ACADEMIES PRESS
Washington, D.C.
www.nap.edu

THE NATIONAL ACADEMIES PRESS   500 Fifth Street, N.W.   Washington, DC 20001

NOTICE: The project that is the subject of this report was approved by the Governing Board of the National Research Council, whose members are drawn from the councils of the National Academy of Sciences, the National Academy of Engineering, and the Institute of Medicine. The members of the panel responsible for the report were chosen for their special competences and with regard for appropriate balance.

Support for this study was provided by the U.S. Geological Survey under Grant Number 07HQAG0124. Any opinions, findings, conclusions, or recommendations expressed in this publication are those of the author(s) and do not necessarily reflect the views of the organizations or agencies that provided support for the project.

This report is available online from the National Academies Press at: http://www.nap.edu.

THE NATIONAL ACADEMIES
Advisers to the Nation on Science, Engineering, and Medicine

The **National Academy of Sciences** is a private, nonprofit, self-perpetuating society of distinguished scholars engaged in scientific and engineering research, dedicated to the furtherance of science and technology and to their use for the general welfare. Upon the authority of the charter granted to it by the Congress in 1863, the Academy has a mandate that requires it to advise the federal government on scientific and technical matters. Dr. Ralph J. Cicerone is president of the National Academy of Sciences.

The **National Academy of Engineering** was established in 1964, under the charter of the National Academy of Sciences, as a parallel organization of outstanding engineers. It is autonomous in its administration and in the selection of its members, sharing with the National Academy of Sciences the responsibility for advising the federal government. The National Academy of Engineering also sponsors engineering programs aimed at meeting national needs, encourages education and research, and recognizes the superior achievements of engineers. Dr. Charles M. Vest is president of the National Academy of Engineering.

The **Institute of Medicine** was established in 1970 by the National Academy of Sciences to secure the services of eminent members of appropriate professions in the examination of policy matters pertaining to the health of the public. The Institute acts under the responsibility given to the National Academy of Sciences by its congressional charter to be an adviser to the federal government and, upon its own initiative, to identify issues of medical care, research, and education. Dr. Harvey V. Fineberg is president of the Institute of Medicine.

The **National Research Council** was organized by the National Academy of Sciences in 1916 to associate the broad community of science and technology with the Academy's purposes of furthering knowledge and advising the federal government. Functioning in accordance with general policies determined by the Academy, the Council has become the principal operating agency of both the National Academy of Sciences and the National Academy of Engineering in providing services to the government, the public, and the scientific and engineering communities. The Council is administered jointly by both Academies and the Institute of Medicine. Dr. Ralph J. Cicerone and Dr. Charles M. Vest are chair and vice chair, respectively, of the National Research Council.

www.national-academies.org

THE NATIONAL ACADEMIES
Advisers to the Nation on Science, Engineering, and Medicine

Water Science and Technology Board
500 Fifth Street, NW
Washington, DC 20001
Phone: 202 334 3422
Fax: 202 334 1961
www.nationalacademies.org/wstb

Dr. Gary L. Rowe
Regional National Water Quality Assessment Program Officer, Central Region
Chair, National Water Quality Assessment Cycle 3 Planning Team
U.S. Geological Survey
Regional Science Office
Denver Federal Center - Building 53, MS 406
W. 6th Avenue and Kipling Street
Denver, CO 80225

Dear Dr. Rowe:

In 2009, the U.S. Geological Survey requested that the National Research Council's (NRC) Water Science and Technology Board review and provide guidance on the direction and priorities of the National Water Quality Assessment (NAWQA) Program. This review would include perspective on past accomplishments and the current and future design and scope of the program as it moves into its third decade of water quality assessment (Cycle 3). In response, the NRC formed the Committee to Review the USGS National Water Quality Assessment (NAWQA) Program in order to address a set of tasks agreed upon by the USGS and NRC (see Attachment B, roster; see Attachment C, Statement of Task).

Once the study was underway, the USGS NAWQA Cycle 3 Planning Team asked the committee to give priority to its first task (see Attachment C) concerning the scientific priorities of the NAWQA program as expressed in its NAWQA Science Framework[1]. The committee was asked to provide an assessment of the Science Framework in terms of whether it sets forth adequately the priorities for the future which will be addressed in the third cycle of the NAWQA program. This letter report provides the committee's response to this request or "guidance on the nature and priorities of current and future water quality issues that will confront the Nation over the next 10-15 years" (see Attachment C, item #1). The committee's final report, anticipated in the spring of 2011, will address the remainder of the first task and the entirety of the statement of task.

The purpose of the Science Framework, the first of two documents on the Cycle 3 design, is "...to outline and describe a framework of water quality issues and priorities for Cycle 3 that reflect the unique capabilities and long term goals of NAWQA, an updated assessment of stakeholder priorities, and an emphasis on identifying potential approaches and partners." It begins with a discussion of NAWQA's unique role in assessing current and future water quality

---

[1] Available online at: http://pubs.usgs.gov/of/2009/1296. The Science Framework is a working document and is the basis for the NAWQA Cycle 3 program.

issues, followed by background on approaches and issues, and concludes with a statement of priority issues facing the nation in the coming years.

NAWQA divided priority issues for Cycle 3 into two categories: 1. Stakeholder Issues Related to Major Environmental *Drivers*, and 2. Stakeholder Issues related to Water Quality *Stressors*. Eleven topical priorities were itemized within the two categories (five drivers, six stressors). Under each priority issue, NAWQA described the nature and scope of the issues articulated by various stakeholders, the program's role and approaches to address the designated issues, and partnerships and collaborative opportunities related to each issue.

Although the Science Framework is a logical and well written document containing an extensive list of water quality issues facing the nation, there remains opportunity for focus and greater clarity. We offer the following suggestions to refocus and reframe the Science Framework. Our intent is to highlight the already achieved and potential scientific impact of the NAWQA program which is critical to future success of NAWQA as it moves into and through Cycle 3. NAWQA is a successful program (NRC 2002, 2009) as it stands and our suggestions are to further improve and help protect what the program already has achieved.

Vision and Principles for NAWQA

From the beginning, a premise of the NAWQA program was a water quality program with *national* impact and coverage. NAWQA's commitment to national level work should be prefaced by a vision for water quality at the national scale. A national water quality program should include national scale surveillance, scenario development, and forecasting. (Scenario development considers how changing land use conditions and climate, for example, may affect water quality in different settings.) It should characterize and evaluate the quality of the nation's waters and serve as a tool for water policy and decision makers in their evaluations of the nation's water resources and their establishment of policies in areas that consider water quality. To this end, the Science Framework, as presented, moves in this direction but needs to be *far more explicit* than implicit in its exposition. The committee recommends that NAWQA better articulate its vision first and foremost in the document and then explicitly describe the value of the program to the nation's water policy and decision makers.

Immediately after presentation of a well articulated vision, NAWQA should outline clarified program principles that are "front loaded" in the planning document. Program principles orient the NAWQA program within the USGS and the federal government. Perhaps most importantly, program principles serve as an internal assessment and guide to keep the program focused and on target. In the following, we highlight program principles and encourage NAWQA to continue this endeavor, making these words their own. We begin with suggesting that the first two program principles address the following points:

- Clearly define and adhere to what *national* means to NAWQA—perhaps to lay down a marker as to where the programmatic tipping point may be from a truly national program to one that lacks adequate spatial coverage and representativeness of conditions to be counted as such. This marker should incorporate consideration of the impacts of abandonment of Study

Units to date, as well as the future of the Study Unit paradigm. NAWQA ought to address the trade offs between benefits from what they plan to "study" and the data given up from the loss of Study Units.

• Identify areas where NAWQA can make a contribution (both social and economic) drawn from research questions and findings that policymakers could expect with respect to water quality.

The committee notes that the Science Framework identifies program principles in Chapter four to guide NAWQA efforts (see Chapter 4: Guiding Principles, Funding Scenarios, and Next Steps for Planning Cycle 3). We support these principles, suggest they follow the two principles mentioned above i.e., are "front loaded" in the document, and recommend a slight expansion in scope (identified below by italics):

• Defining NAWQA's role, or scientific *areas where NAWQA can make unique and substantial contributions such as monitoring for nutrients or sediment,* in water quality assessment,
• Develop NAWQA priorities to be consistent with the six recently designated USGS Strategic Science Directions,
• Maintain continuity of long term goals and design of the program, *i.e., status, trends, and understanding.*

Water Quality Drivers and Stressors

The USGS is commonly viewed as an independent, unbiased, non regulatory driven and high quality source of data and its interpretation. Indeed, the USGS Water Resource Discipline was encouraged by the NRC Committee on USGS Water Resources Research to "lead the nation in water science" (NRC, 2009). NAWQA has the ability to illuminate and address national water quality issues and we encourage the program to do this, within its purview. Moreover, with a national scope NAWQA is well placed to address big picture drivers or causes of change and issues related to water quality that the nation faces. Translating this to the Science Framework, we recommend NAWQA reframe the planning document around big picture drivers.

We do not find the major environmental drivers and water quality stressors itemized in the Science Framework mutually exclusive. The terms "driver" and "stressor" are linked. We consider drivers regional or national scale anthropogenic and natural forces that directly or indirectly cause stress, or changes, to water supplies and associated ecosystems at multiple scales. For example, one *driver* would be climate change, which causes *stressors* such as increased storm intensity. In other words stressors constitute technical topics, or "priorities" that should be structured within the context of the drivers, or "causes", for changing water quality and key policy relevant questions that NAWQA hopes to answer.

Specifically, we recommend that NAWQA reorganize its activities to focus on the two major large scale drivers affecting national water quality: (1) change in land use due to population and other demographic changes; and (2) climate variability and change, which were mentioned in the USGS Science Framework, although not in this context. As such, these drivers are clearly important to both NAWQA and its stakeholders. The committee and others (NRC, 2009) agree with this importance and note that a large majority, if not all, of the stressors on aquatic systems and changing water quality link directly to a changing climate and changing land use practices. These large scale drivers provide a fully adequate umbrella under which most environmental challenges and, ultimately, the NAWQA priorities can be identified.

We define land use change due to population and other demographic changes as change in the use of land for cities, for agriculture (including changes in crop type), for forestry, etc. due to multidimensional changes in numbers of people, their geographical distribution, the age distribution, and other changes over time. Such changes generate an evolving alteration of the landscape and impacts on water quality. For example, large-scale agricultural practices changed the landscape in the Midwest such that it now stresses the Mississippi River system and Gulf of Mexico with excessive nutrient runoff and sediment influx from erosion. The same stressors are also the result of expanded transportation networks based on cars and increased impervious surfaces as land use changes from urban to suburban or exurban areas.

Climate variability and change drives many stressors related to water quality. This includes altering the balance within the hydrologic cycle, which impacts infiltration and recharge to watersheds, aquifers, and river base flows, precipitation frequency and intensity, flooding and storm surge, and water storage in snowpacks. For example, in the Western U.S., mountain snowpack is diminishing and melting earlier than in the past although the total volume of precipitation is not changing significantly. This is 1) increasing late winter and early spring runoff and 2) reducing spring and summer runoff. The latter will impact water quality by producing higher stream temperatures and concomitant lower dissolved oxygen levels along with less water for waste dilution, whereas the former may increase flooding with associated increases in sediment and contaminant loads (Service, 2004). Also, higher water temperatures earlier in the season combined with nutrient wash off in early spring through melt or rain will likely lead to increased algal blooms and eutrophication frequency.

In the face of (water resource) challenges caused by climate and land change, policy "Decisions will be made, with or without scientific input." (NRC, 2009), and logically using science best meets the needs of society. To that end, NAWQA managers and scientists need to think about which components of these two major drivers they best can tackle. The other nine "drivers" and "stressors" in the Science Framework are subtopics that can be addressed under these major items.

The committee recommends NAWQA explicitly lay out *policy relevant research questions* under the auspices of each driver. These research questions will convey to decision makers and water managers the important topics that the NAWQA program will address as well as the critical value of the NAWQA program itself. An example of a policy relevant question might be: How would changing land use and a changing climate affect water quality, quantity, and allocation in the American west? Or, how will changing climate and land use affect the balance of human water needs and valued ecosystem needs in different regions of the United States?

To do as we suggest, NAWQA leadership should first determine how it can use the program and other historic data and the USGS forecasting and scenario development abilities to answer policy relevant research questions that demonstrate program impact. The answers to this determination will help rank program priorities that consider the two major water quality drivers, change in land use due to population and demographic change and climate change, facing the nation. As an example, consider how Midwestern agriculture has led to hypoxia in the Gulf of Mexico. The driver in this case is *change in the intensity of agricultural land use due to cropping practices, in part, related to population and demographics,* and the stressors are *sediment* and *dissolved nutrients.* The policy relevant research question is addressing how *future change in agricultural practices would aid in remediation of Gulf Coast hypoxia?* and the impact is a contribution to one of the most challenging water quality issues facing the nation.

A Case for Clarity

We cannot emphasize enough that NAWQA should be clearer with respect to its purpose. The committee finds refocusing and reframing of the document to be the first step to clarity. This will elucidate exactly which program priorities from the Science Framework will best serve a nation facing significant water quality challenges related to changing land use and a changing climate. Taking this a step further, NAWQA should define the *scale* of endeavors, articulate *specific examples* of activities, and define key terms (e.g. "water quality", "ecosystem health", "microbial contaminants") within each priority.

The lack of attention to scale in the Science Framework was surprising. Scale is *a* if not *the* key component that makes NAWQA a unique program. Again, a national water quality program should include *national scale* surveillance, scenario development, and forecasting. It should characterize the quality of the nation's waters and serve as a tool for water policy and decision makers in their evaluations of the nation's water resources. Scale constitutes a unique niche for NAWQA, compared to science done by other federal agencies. Many components of water problems are local to regional in nature. Yet scale issues emerge naturally as the interaction between land use change due to population and other demographic changes, and climate change is considered for different water quality attributes. NAWQA has successfully linked the regional and local variations into national synthesis of water quality assessments. NAWQA should continue to consider how processes in regional studies are linked on a national scale. Furthermore, as the program moves into Cycle 3, scale should be considered on each level of the planning document from the articulation of the program vision to its role in regional and topical studies as well as the sampling and modeling design for both assessment and prediction.

The generic approach to the partnership sections was uninformative; NAWQA should be more explicit in its plan to execute collaboration. We define collaboration as working together, publishing together, and leveraging resources or science capabilities (data, human resources, modeling capacity, etc.). Appropriate clarification would include specific examples of current and future collaborative partners, programs both inside and outside the U.S. (e.g., various EPA programs, relevant Canadian and Mexican activities, and NSF's planned NEON, STREON, and WATERS), and activities with particular attention to how NAWQA fits with the new USGS Water Census initiative as well as how water resources are connected to other countries.

In the spirit of achieving clarity, we provide specific comments on each of the Science Framework's eleven priorities in the context of our suggested reframing of the document. Our criteria for these comments are that NAWQA should focus on national issues, should combine priority areas when possible, and should concentrate on areas where NAWQA can make unique and substantial contribution.

First and foremost, Policies, Regulations, and Management Practices and Effects of Multiple Stressors are cross cutting topics that should be considered and integrated into all programmatic activities. They are design principles, not priorities, that are unevenly plugged in across the document. Rather, all priorities should be defined in terms of policy needs with consideration of multiple stressors. For example, what does a national synthesis report on topic "X" mean with respect to policy relevant topics currently being considered? What are the big picture questions answered by such a synthesis that capture the implications of multiple stressors on water quality?

Addressing Common Chemical Contaminants is NAWQA's "bread and butter", and obviously should remain a core priority. NAWQA should carefully consider Microbial Contaminants within the scope of a national vision. It is nonetheless critical that NAWQA articulate policy relevant research questions that connect these contaminant issues to climate change and land use change and population growth in order to clarify NAWQA's approach and show the relevance of its work within these priority areas. Terms such as "limit human use" or "affect aquatic ecosystem health" appear frequently and require definition and clarification. How can these terms be assessed using field measurements? Particular attention should be paid to how NAWQA coordinates and collaborates with EPA in the context of drinking water.

The committee supports NAWQA continuing its work on eutrophication. Yet eutrophication, as a priority, seems no different than understanding the results of monitoring of nutrients and/or other parameters such as chlorophyll a. Understanding the process of eutrophication is an outcome of NAWQA's monitoring for Common Chemical and Microbial Contaminants. Therefore, we recommend eutrophication be subsumed as a component of Common Chemical and Microbial Contaminants and related to the two major drivers—as the drivers are "source terms" affecting nutrient loading that result in eutrophication. It should be clear that NAWQA's work on eutrophication falls within the scale of major river and estuarine environments when considering nutrient related policy questions. It is important to define NAWQA's partnership role thoroughly here, particularly with entities outside the agency (e.g., with NOAA and EPA).

Sediment is a critical issue that NAWQA is well positioned to address. How the sediment delivered in response to changing land use influences aquatic ecosystems and how sediment may be controlled within dam management and operations while minimizing ecological and economic impact constitute examples of policy relevant issues. Detailed sediment flux and discharge monitoring is very costly and may be beyond NAWQA's means, but NAWQA has the capacity to address sediment in the context of key questions aimed at addressing major ecosystem and economic impacts using SPARROW modeling (SPAtially Referenced Regressions On Watershed attributes) as a central tool. NAWQA should bring to bear its hydrologic and geologic expertise on this issue to complement and enhance the engineering perspective driven by other federal agencies.

Wastewater Reuse does not appear within the NAWQA purview because of scale and should be omitted as a high priority. While there are clear water quality issues related to reuse, most projects are local in nature and would not be well suited for integration with the larger national priorities that NAWQA should address. (Concrete examples of how the NAWQA program addresses this issue on a national scale would be necessary prior to further pursuit.) Wastewater Reuse would seem a topic that the new USGS Water Census initiative might tackle. NAWQA certainly needs to stay abreast of Water Census developments, and perhaps work collaboratively with the Water Census.

Hydrologic Modification and Flow Modification should be merged and considered in tandem. By adopting this approach, NAWQA planning would unify all activities probing flow and hydrologic modification, including drinking water. Policy relevant questions within this priority could overlap with Sediment, i.e., questions regarding flow modification and sediment flux. Clarification of priority scope is necessary to distinguish between Sediment and Hydrologic Modification and Flow Modification.

NAWQA should play a careful role with respect to Emerging Contaminants to avoid getting caught up in the "contaminant of the day." First, NAWQA needs to clearly define the scientific concerns with respect to this issue; why should certain emerging contaminants or contaminant classes be monitored? As part of this effort, NAWQA needs to clearly address how it defines "emerging contaminants". The term represents a huge continuum of compounds that makes a "one size fits all" approach inappropriate and, frankly, intractable. If scientific concern is deemed adequate, NAWQA should move into contaminant areas for which there are clear, established methods and approaches or in which it can do meaningful surveillance. We suggest NAWQA begin with only special projects on emerging contaminants or those driven by clear scientific concern and thus, policy considerations. Careful attention should be paid to coordination with the USGS Toxics Program whose mission is to conduct field based research to understand behavior of toxic substances in the nation's hydrologic environments in support of the development of strategies to clean up and protect water quality.

NAWQA can address only some portions of Energy and Natural Resource Development priority. Some issues within this priority are too localized for NAWQA to address. NAWQA leadership should think through what they can do well in this arena and the resulting strategy might be segmented. For example, the NRC, in its Cycle 2 NRC report (NRC, 2002), suggested that NAWQA leadership should evaluate clearly whether it actually had the resources to comprehensively address water quality degradation related to mining. NAWQA is best positioned in Energy and Natural Resources Development with respect to biofuel development which can be addressed under both drivers. The committee commends and encourages this work. However, NAWQA does not seem well positioned for evaluating water degradation caused by energy development overall. The committee does not advise NAWQA to take the lead on this issue among the agencies that deal with water resources or within the USGS.

Summary

The Science Framework is an opportunity to demonstrate the past, present, and future impacts of NAWQA and to articulate a compelling case for the need for NAWQA—a need in which the committee strongly believes. NAWQA is a unique program within a unique agency filling the niche of producing high quality national water quality data and interpretation; it is unequaled by any other entity. The committee urges creation of a more focused, restructured, and clarified planning document for Cycle 3 of the NAWQA program. It should clearly and compellingly demonstrate how the program has had and will have an impact on national water policy, and, in part, secure that NAWQA moves through Cycle 3 intact as our nation's premier water quality monitoring program.

Sincerely,

Donald I. Siegel, *Chair*
Committee to Review the USGS
National Water Quality Assessment
(NAWQA) Program

Attachment A: References
Attachment B: Committee Membership
Attachment C: Statement of Task
Attachment D: Acknowledgement of Reviewers

cc:
Matthew Larsen
Donna Myers

ATTACHMENT A

REFERENCES

National Research Council (NRC). 2002. Opportunities to Improve the U.S. Geological Survey National Water Quality Assessment Program. Washington, D.C.: National Academies Press.

National Research Council (NRC). 2009. Towards a Sustainable and Secure Water Future: A Leadership Role for the USGS. Washington, D.C.: National Academies Press.

Service, R. F. 2004. As the west goes dry. Science 303(5661):1124-1127.

ATTACHMENT B

**COMMITTEE ON PREPARING FOR THE THIRD DECADE (CYCLE 3) OF THE
NATIONAL WATER QUALITY ASSESSMENT PROGRAM**

Donald I. Siegel, *Chair*, Syracuse University
Michael E. Campana, Oregon State University
Jennifer A. Field, Oregon State University
George R. Hallberg, The Cadmus Group, Inc.
Nancy K. Kim, State of New York Department of Health
Debra S. Knopman, RAND Corporation
Upmanu Lall, Columbia University
Walter R. Lynn, Cornell University
Judith L. Meyer, University of Georgia
David W. Schindler, NAS, University of Alberta
Deborah L. Swackhamer, University of Minnesota

NRC Staff
Laura J. Helsabeck, Project Director
Anita Hall, Project Assistant

ATTACHMENT C

STATEMENT OF TASK

The project will provide guidance to the U.S Geological Survey on the design and scope of the NAWQA program as it enters its third decade of water quality assessments. T he committee will assess accomplishments of the NAWQA program since its inception in 1991 by engaging in discussions with the Cycle 3 Planning Team, program scientists and managers, and external stakeholders and users of NAWQA data and scientific information. The committee will also review USGS internal reports on NAWQA's current design for monitoring, assessments, research, and relevance to key water topics. The main activities of the study committee will be to:

1. Provide guidance on the nature and priorities of current and future water quality issues that will confront the Nation over the next 10-15 years and address the following questions:
   - Which issues are currently being addressed by NAWQA and how might the present design and associated assessments for addressing these issues be improved?
   - Are there issues not currently being substantially addressed by NAWQA that should be considered for addition to the scope of NAWQA?

2. Provide advice on how NAWQA should approach these issues in Cycle 3 with respect to the following questions:
   - What components of the Program—Surface Water Status and Trends; Ground-Water Status and Trends; Topical Understanding Studies; National Synthesis— should be retained or enhanced to better address national water quality issues?
   - What components of the program should change to improve how priority issues are addressed?
   - Are there new Program components that should be added to NAWQA to enable the Program to better address and analyze National water quality issues and related public policy issues?

3. Identify and assess opportunities for the NAWQA Program to better collaborate with other Federal, State, and local government, non-governmental organizations, private industry, and academic stakeholders to assess the nation's current and emerging water quality issues.

4. Review strategic science and implementation plans for Cycle 3 for technical soundness and ability to meet stated objectives.

ATTACHMENT D

ACKNOWLEDGMENT OF REVIEWERS

This letter report has been reviewed in draft form by individuals chosen for their diverse perspectives and technical expertise, in accordance with procedures approved by the National Research Council's Report Review Committee. The purpose of this independent review is to provide candid and critical comments that will assist the institution in making its published report as sound as possible and to ensure that the report meets institutional standards for objectivity, evidence, and responsiveness to the study charge. The review comments and draft manuscript remain confidential to protect the integrity of the deliberative process.

We wish to thank the following individuals for their review of this report: Kenneth R. Bradbury, University of Wisconsin and the Wisconsin Geological and Natural History Survey; Joan G. Ehrenfeld, Rutgers University; Mike Kavanaugh, Malcolm Pirnie, Inc.; Kenneth H. Reckhow, Duke University; and Marylynn Yates, University of California, Riverside.

Although the reviewers listed above have provided many constructive comments and suggestions, they were not asked to endorse the conclusions or recommendations nor did they see the final draft of the report before its release. The review of this report was overseen by Henry J. Vaux, Jr., University of California, Berkeley. Appointed by the National Research Council, he was responsible for making certain that an independent examination of this report was carried out in accordance with institutional procedures and that all review comments were carefully considered. Responsibility for the final content of this report rests entirely with the authoring committee and the institution.

Appendix B

Letter Report Assessing the USGS National Water Quality Assessment Program's Science Plan

LETTER REPORT ASSESSING THE USGS NATIONAL WATER QUALITY ASSESSMENT PROGRAM'S SCIENCE PLAN

Committee on Preparing for the Third Decade (Cycle 3)
of the National Water Quality Assessment (NAWQA) Program

Water Science and Technology Board

Division on Earth and Life Studies

NATIONAL RESEARCH COUNCIL
OF THE NATIONAL ACADEMIES

THE NATIONAL ACADEMIES PRESS
Washington, D.C.
www.nap.edu

THE NATIONAL ACADEMIES PRESS   **500 Fifth Street, N.W.   Washington, DC 20001**

NOTICE: The project that is the subject of this report was approved by the Governing Board of the National Research Council, whose members are drawn from the councils of the National Academy of Sciences, the National Academy of Engineering, and the Institute of Medicine. The members of the panel responsible for the report were chosen for their special competences and with regard for appropriate balance.

Support for this study was provided by the U.S. Geological Survey under Grant Number 07HQAG0124. Any opinions, findings, conclusions, or recommendations expressed in this publication are those of the author(s) and do not necessarily reflect the views of the organizations or agencies that provided support for the project.

This report is available online from the National Academies Press at: http://www.nap.edu.

Printed in the United States of America.

THE NATIONAL ACADEMIES
Advisers to the Nation on Science, Engineering, and Medicine

The **National Academy of Sciences** is a private, nonprofit, self-perpetuating society of distinguished scholars engaged in scientific and engineering research, dedicated to the furtherance of science and technology and to their use for the general welfare. Upon the authority of the charter granted to it by the Congress in 1863, the Academy has a mandate that requires it to advise the federal government on scientific and technical matters. Dr. Ralph J. Cicerone is president of the National Academy of Sciences.

The **National Academy of Engineering** was established in 1964, under the charter of the National Academy of Sciences, as a parallel organization of outstanding engineers. It is autonomous in its administration and in the selection of its members, sharing with the National Academy of Sciences the responsibility for advising the federal government. The National Academy of Engineering also sponsors engineering programs aimed at meeting national needs, encourages education and research, and recognizes the superior achievements of engineers. Dr. Charles M. Vest is president of the National Academy of Engineering.

The **Institute of Medicine** was established in 1970 by the National Academy of Sciences to secure the services of eminent members of appropriate professions in the examination of policy matters pertaining to the health of the public. The Institute acts under the responsibility given to the National Academy of Sciences by its congressional charter to be an adviser to the federal government and, upon its own initiative, to identify issues of medical care, research, and education. Dr. Harvey V. Fineberg is president of the Institute of Medicine.

The **National Research Council** was organized by the National Academy of Sciences in 1916 to associate the broad community of science and technology with the Academy's purposes of furthering knowledge and advising the federal government. Functioning in accordance with general policies determined by the Academy, the Council has become the principal operating agency of both the National Academy of Sciences and the National Academy of Engineering in providing services to the government, the public, and the scientific and engineering communities. The Council is administered jointly by both Academies and the Institute of Medicine. Dr. Ralph J. Cicerone and Dr. Charles M. Vest are chair and vice chair, respectively, of the National Research Council.

www.national-academies.org

THE NATIONAL ACADEMIES
Advisers to the Nation on Science, Engineering, and Medicine

Water Science and Technology Board
500 Fifth Street, NW
Washington, DC 20001
Phone: 202 334 3422
Fax: 202 334 1961
www.nationalacademies.org/wstb

Dr. Marcia McNutt
Director, U.S. Geological Survey
USGS National Center
12201 Sunrise Valley Drive
Reston, VA 20192, USA

Dear Dr. McNutt,

In 2009, the U.S. Geological Survey requested that the National Research Council's (NRC) Water Science and Technology Board review and provide guidance on the direction and priorities of the National Water Quality Assessment (NAWQA) Program. This review would include perspective on past accomplishments and the current and future design and scope of the program as it moves into its third decade of water quality assessment (Cycle 3). In response, the NRC formed the Committee to Review the USGS National Water Quality Assessment (NAWQA) Program to address a set of tasks agreed upon by the USGS and NRC (see attachment B, roster; see attachment C, statement of task). The NRC's Water Science and Technology Board has a history of advising the NAWQA Program since its conception in the mid-1980s. This committee has continued that advisory role authoring a letter report on the initial Cycle 3 planning document, the Science Framework (*Letter Report Assessing the USGS National Water Quality Assessment Program's Science Framework* (NRC, 2010)). Based on advice contained in that letter report, input from stakeholders, and additional reflection from the NAWQA Cycle 3 Planning Team, the Science Framework evolved into the Cycle 3 Science Plan[1]. The Science Plan is the high level planning document that will guide the NAWQA program through the next 10 years of water quality monitoring.

Your letter dated December 14th, 2010 asked the committee to provide additional advice on NAWQA's progress in the Cycle 3 planning process, focusing particularly on whether the draft NAWQA Science Plan sets forth adequate priorities and direction for the future. We are responding to your request through this letter report, which partly addresses our tasks 1 & 4 (see attachment C) to provide guidance "on the nature and priorities of current and future water quality issues facing the nation" and "to review strategic science and implementation plans for Cycle 3 for technical soundness and ability to meet stated objectives." Our committee's final report, anticipated in the summer of 2011, will address the entirety of the statement of task.

The Science Plan

Over the past 20 years, the nation has invested in the NAWQA program to probe the status of, trends in, and understanding of the nation's water quality. This investment in NAWQA has resulted in methodological advances (e.g., national sampling protocols, analytical methods, groundwater field investigative tools), conceptual and intellectual advances such as the development and implementation of predictive tools (e.g., models), and national syntheses of critical water quality topics. Now, NAWQA is the nationally-recognized program responsible for evaluating the nation's water quality. To continue this evaluation into its third decade, the NAWQA Cycle 3 Science Plan contains four goals: 1) *Data Collection and Trend Assessment*, 2) *Interpretation and Understanding* of these data relative to land use and climate variability; 3) targeted studies for the *Determination of the Cause-Effect Relationships of Multiple Stressors and Multiple Effects*; and 4) using these data, understanding, and relationships to *Forecast Future Trends* of pollutants under different scenarios of land use, climate, and resource management.

NAWQA is poised, both within the USGS and the federal government, to continue the requisite sampling of our nation's waters (Goal 1) to understand the interplay between the complex factors that affect water quality (Goal 2). The committee supports the continuation of these priorities including the choice of four major stressors (contaminants, streamflow alteration, nutrients, and sediment). Yet NAWQA is now also in a position to produce an even larger payoff. The program has reached a threshold in which the value of achieving Goals 3 (effects of stressors) and 4 (forecasting) is greater than that achieved by the sum of its parts. In other words, NAWQA has evolved from a water quality program emphasizing data collection and trend assessments to one that has the potential to predict and forecast pollutant occurrence and trends under multiple scenarios at nationally significant scales. The program's scientific investments are maturing, enabling NAWQA to move past the current water quality monitoring to understanding the dynamics of water quality changes and using that understanding to forecast likely future conditions. By building on and maintaining the foundation from Cycle 1 and Cycle 2, NAWQA should move into the arena of "dynamic water quality monitoring" (Box 1). These are advances that the nation needs and the committee strongly supports.

Box 1
Traditional Water Quality Monitoring vs. Dynamic Water Quality Monitoring

In Cycle 1 and 2, NAWQA assessed the status and trends of the nation's water quality through a "Traditional Water Quality Monitoring" approach or by collecting data at regular intervals using a combination of fixed site and rotational sampling strategies. A "Dynamic Water Quality Monitoring" approach would assess the dynamics of water quality changes in addition to status and trends by a sampling design adaptable in both frequency and location overlaid on the traditional fixed sampling strategy. For example, changing sampling frequency to capture the dynamics of wet or dry spells associated with El Nino/La Nina events. By selectively increasing temporal and spatial resolution when and where it is needed, dynamic monitoring contributes to understanding of complex water quality phenomena and allows improved forecasting of likely future conditions.

The committee compliments the NAWQA Cycle 3 Team for envisioning a bold plan for the coming decade, with priority placed on dynamic water quality monitoring. Also, this version of the Science Plan responded to the comments made by the committee in the first letter report (NRC, 2010). Yet the Science Plan needs to continue to improve its clarity and NAWQA should continue to enhance the effectiveness of its communication of the ideas noted above. Although explaining the importance of the Cycle 3 goals and how NAWQA intends to accomplish these goals is essential, it is also critical to explain *why* these goals need to be pursued. Specifically, the plan needs to clarify "Why the USGS?", "Why now?", and "Why NAWQA?" much like it was presented to the committee in the fall of 2010 (October 26th, 2010 open session meeting of this committee). Why dynamic water quality monitoring is important now, and why the USGS via NAWQA can achieve this needs further clarity in the document although the concept and the need is compelling. Including points such as the following will enhance the draft Science Plan:

- Simply maintaining traditional water quality monitoring will result in USGS lagging behind in providing the necessary science to solve the nation's water problems as population growth, changes in land use, and climate variability continue to stress our nation's water resources;
- Water resources problems need to be addressed through a systems approach by considering a range of effects on water quality caused by multiple stressors;
- NAWQA is uniquely positioned to lead the nation in a dynamic national synthesis of water quality information and understanding because it has infrastructure in place, interdisciplinary and collaborative experience, state-of-the-art analytical capability, and modeling capacity to do this work (NRC 2002; NRC, 2009);
- NAWQA provides unique management-relevant assessments and tools within the public domain and has developed the capability and coordination to get needed science to decision makers (USGS, 2010);
- NAWQA Cycle 3 and the corresponding Science Plan are an excellent investment for the nation because Goals 3 and 4 provide considerable added value and logically evolve from the work proposed in Goals 1 and 2.

Outputs and Potential Outcomes

NAWQA's Science Plan has four goals, with objectives under those goals. The Science Plan should identify key expected outputs (the products) and potential outcomes from each objective. Outputs and potential outcomes are identified for the objectives under Goals 1 and 2. Outputs and potential outcomes are described under Goal 3, but are not objective specific. Outputs and potential outcomes are not provided for Goal 4. Developing outputs and potential outcomes for each goal is viewed as critical for the science plan's implementation, to help frame the significance of dynamic water quality monitoring, and to help NAWQA allocate its resources effectively and efficiently over the next 10 years. To the extent possible, NAWQA should estimate when the potential outcomes are expected to occur. (The committee acknowledges that what is practical within a research-oriented product approach may be different from what is needed in a public information-oriented product approach.) Description of deliverables and their timing will help NAWQA implement its Science Plan and help its partners and stakeholders plan how and when they will utilize NAWQA's work. USGS should strive to make NAWQA data, synthesis, and model projections available to users as quickly as practical, increasing the usefulness and relevance of its work.

Trends vs. Dynamics

Traditional monitoring assesses change by periodic measurements (for example, in the same seasons) to establish baseline water quality attributes and their seasonal averages. Results from regular sampling in time can help identify periodic changes in the state of the system with some recognition of climate or other changes in water quality, but cannot lead to a more fine-tuned understanding of trends and why change occurs. For example, the cumulative effects of changes in climate on the spatial-temporal attributes of water quality and ecosystem response may be sudden and dramatic (Lipp et al., 2001). A wet period may be marked by greater than average frequency and intensity of sediment entrainment and transport, leading to higher nutrient, pesticide and pathogen loadings into a receiving water body. A dry or quiescent period would be marked by the subsequent biogeochemical transformation of these loads in the water-soil columns. A dynamic sampling strategy designed to capture specific events and changes and not designed to follow a strict periodicity, would be able to contribute to understanding the relationships of variable and multiple stressors and their effects.

As NAWQA moves forward with a more dynamic approach to its program, the distinction between sampling parameters for traditional water quality monitoring and sampling for dynamic water quality changes becomes more important. NAWQA has utilized a periodic approach in assessments of pesticides in hydrologic systems and found remarkable added value (Box 2). NAWQA leaders should continue to recognize that aquatic systems constantly fluctuate, rather than assume they operate uniformly such that sampling can be done only in a uniform way. As such, the NAWQA monitoring and modeling design should reflect a dynamic sampling strategy overlain on top of a periodic sampling design (Box 1 and 2). The dynamic part of the sampling design would be question based, supporting Goals 2, 3, and 4, whereas the traditional design maintains documenting long term trends in water quality (Goal 1). This pairing provides an opportunity for innovation through an adaptive monitoring system that follows some of the key questions in the Science Plan.

Box 2
The Importance of Sampling for Dynamic Water Quality Monitoring

Beginning in the mid-1990s, NAWQA collected samples and probed the presence of the insecticide diazinon in an urban stream. Diazinon samples were collected yearly, rather than the 4 year rotational sampling design commonly employed by NAWQA, during Cycle 2. NAWQA continued sampling as diazinon was phased out both in indoor and outdoor residential use in the early 2000s. NAWQA developed a reliable time-series model for assessing long term changes in diazinon concentrations as residential use declined. The model showed a rapid water quality response to eliminating outdoor uses in 2002 and a continued decline in diazinon concentration through 2004. Furthermore, NAWQA examined the results as if the 4 year rotational sampling design was employed, i.e., if the model was based on sampling every 4th year. The resulting trend indicated an *increase* in diazinon through 2004, rather than the decrease in concentration that actually occurred as a result of phasing out use of the insecticide.

SOURCE: October 26th, 2010, personal communication, Robert J. Gilliom.

Ecosystem Services

Aquatic ecosystems both impact and are impacted by water quality (NRC, 1995; NRC, 1992). By focusing on how water quality impacts ecosystems, the Science Plan addresses only half of the picture. Consequently, aquatic ecosystems only appear subjected to degraded water quality. The Science Plan should recognize that biogeochemical processes in aquatic ecosystems also condition the water quality in those ecosystems or explain that the biogeochemical processes that are characteristic of aquatic ecosystems in good condition help restore and maintain water quality, i.e., there are feedback loops in the system. In addition, the Science Plan presents human and ecosystem needs for water as though they are two separate issues. In fact, meeting ecosystem needs for water ensures the maintenance of biogeochemical processes that result in high quality water for humans. To be clear, the letter report is only suggesting that NAWQA acknowledge these feedbacks and synergy in the Science Plan, not to change priorities as they are currently listed.

Linking Groundwater and Surface Water

The NAWQA Program has progressed greatly in its understanding and simulation of surface water - groundwater interactions. The initial Cycle 1 study unit design in the late 1980s specified 69 surface water study units and 54 groundwater study units with little consideration for the interconnection between surface water and groundwater within a given study unit. With advice from the NRCs *Committee to Review the USGS National Water Quality Assessment Pilot Program* (NRC, 1990), the USGS adopted a more integrated approach with respect to surface water and groundwater interaction by implementing 60 "integrated" study units in Cycle 1. (As a result of budget cuts, the number was reduced to 42 and eventually phased out in Cycle 2.) It is proposed to replace the study unit design in Cycle 3 with Integrated Watershed Studies (IWS) for surface water and the Principal Aquifer (PA) as the organizing unit for groundwater. Although surface water and groundwater are not segregated as in the original (pre-1990) NAWQA concept in the Science Plan, NAWQA should remain vigilant to ensure the proper characterization of surface water and groundwater interactions and their effects on water quality within the new design.

Sediment

Excess sediments and turbidity are among the top ten causes of impairment in U.S. rivers and streams (EPA, 2009). The inclusion of measures of sediment transport and impacts in Cycle 3 is a much needed addition to the NAWQA program. In its review of plans for Cycle 2, NRC recommended the inclusion of sediments, recognizing that USGS was the federal agency with unique expertise to tackle this problem (NRC, 2002). Budget constraints prevented this addition from happening in Cycle 2. Taking advantage of technological innovations, in Cycle 3 NAWQA now proposes to use surrogate measures (for example, optical backscatter or acoustic sensors) to develop estimates of sediment transport using statistical software. This approach promises to be an efficient way to provide valuable water quality information (Gartner, 2002; Gartner, 2004; Gartner et al., 2001; Thorne and Hanes, 2002). Furthermore, coupling sediment characterization in river systems with the SPARROW model offers considerable promise for management applications. On a smaller scale, SPARROW has been used to assess where management interventions would be most effective in reducing sediment transport in Chesapeake Bay watersheds (Brakebill et al., 2010). Incorporation of sediment measures in Cycle 3 offers the promise of this kind of application for priority-setting at multiple and larger scales. The Science Plan would be improved if these applications were more clearly articulated.

NAWQA's Value in a Reorganized USGS

To enhance the work of the agency, the USGS is currently realigning its leadership and budget structure around interdisciplinary themes or mission areas related to the science strategy "Facing Tomorrow's Challenges—U.S. Geological Survey in the Decade 2007-2017" (UGSG, 2007). The 2009 NRC report, *Towards a Sustainable and Secure Water Future*, pointed out that critical water-related issues occur within most if not all new USGS Science Strategy mission areas (NRC, 2009). NAWQA is well positioned to contribute to these mission areas, building on its success in multidisciplinary efforts within the USGS over the last few decades (Box 3), but this is not well articulated in the Science Plan. A discussion in the spirit of and building from the examples listed in the following paragraph should be articulated in the Science Plan.

Box 3

"...every preceding chapter of this report notes examples of cooperative efforts. In the committee's view, NAWQA program staff have done an excellent job of establishing cooperative relationships within the USGS and external programs. These efforts have strengthened NAWQA and have improved the viability and visibility of the USGS as a whole."

SOURCE: Chapter 7 of *Opportunities to Improve the U.S. Geological Survey National Water Quality Assessment Program* (NRC, 2002).

A continued relationship between NAWQA and programs in the *Ecosystems Mission Area* would be valuable to the USGS. NAWQA has integrated ecological components with physical and chemical measurements with the co-location of ecological and water quality sampling sites (NRC, 2009). NAWQA science has enhanced understanding of the effects of urbanization, mercury, and nutrients on stream ecosystems through Topical Studies in Cycle 2. NAWQA is currently developing a "data warehouse" for biological information, in collaboration with other disciplines and programs within the USGS. NAWQA and the Toxic Substances Hydrology program (now part of the *Energy and Minerals, and Environmental Health Mission Areas*) have a long history of successful, joint collaboration (NRC, 2009; NRC, 2002). The USGS leads the way in identification, tracking, and doing research on emerging contaminants, a role resulting in part from collaboration between the USGS Toxic Substances Hydrology Program and NAWQA (Kolpin et al., 2002). A NAWQA-Toxics effort produced a set of three papers on Mercury Cycling in Stream Ecosystems that were published in the April 15th, 2009 issue of *Environmental Science and Technology* and are one of the most comprehensive studies of stream mercury dynamics. One of NAWQA's noted accomplishments has been the linkage of land-use to water quality conditions. In Cycle 3, NAWQA proposes enhancing its consideration of climate change issues and water. This could be particularly valuable to and invite important collaborative opportunities with the *Climate and Land-Use Change Mission Area*. And certainly, NAWQA's long-standing work in data integration, as well as its experience developing a data warehouse to provide accessible data to other agencies and the public, is relevant to the work of the *Core Science Systems* mission.

NAWQA program leaders should seek further opportunities for collaboration within the agency. For example, in the early days of NAWQA the program pioneered internal capabilities for database management, communications, and external coordination to meet program needs that were either not available or were insufficiently developed within the Water Resources Division or the USGS at that time. Since then, the USGS has developed some of these services and resources more fully and offers support to all programs within the USGS. After 20 years of NAWQA operations in parallel with these significant changes in USGS capabilities, particularly in the USGS Office of Communications, the committee sees value in NAWQA management revisiting the relative merits of using NAWQA program funds to handle communications and possibly other program support services instead of drawing on comparable services and resources provided at the agency level.

As noted, NAWQA has a history of working in the multidisciplinary, collaborative interface and could serve as a useful resource and model to assist in the realignment of the agency to multidisciplinary and cross-disciplinary missions. Although defining collaboration and listing partners is important to

NAWQA planning efforts, true collaboration begins with identifying common questions or goals shared with other mission areas and USGS programs. To be effective in this effort, NAWQA must more clearly identify how its goals are linked to the newly formed USGS mission areas framed from themes in the USGS Science Strategy. NAWQA should make a systematic effort to communicate its capabilities and potential value to the relevant programs and offices within the USGS through the Science Plan. These communications are a two-way street, opening up the possibility of improved coordination within the USGS and potentially greater use of NAWQA data and analysis by the other program areas. Furthermore, fiscal realities highlight the need to seize these collaborative opportunities within the USGS and the re-organization is a window of opportunity for this to be fully realized.

Conclusion

The NAWQA program has matured over its two decades and is at a point where it should not simply continue its previous work but should do the dynamic water quality monitoring that is proposed for Cycle 3. This is a compelling plan for the program that the committee strongly supports; in Cycle 3 NAWQA will advance the understanding of the dynamics of water quality change and forecast likely future conditions. The committee supports the Cycle 3 priority of dynamic water quality monitoring. The Science Plan is technically sound and the NAWQA program has the scientific capability to achieve the Science Plan objectives. Yet the concept of dynamic water quality monitoring needs further development in the Science Plan. For example, a strong justification for why dynamic water quality monitoring is important, why now and why the USGS via NAWQA can achieve this remains unwritten. Further defining program outputs and potential outcomes will also help frame the significance of dynamic water quality monitoring. Moreover, thinking through a dynamic, question-driven sampling strategy to execute this concept will serve the program well.

The committee stresses that the NAWQA assessment of the nation's water quality through the long term benchmark data collection should not be discarded because of program movement towards dynamic water quality monitoring. However, the need to not just collect data at regular snapshots in time and document trends, but to also capture attributes of the events that define the baseline trends so that we have not just trend identification but also an attribution aspect as part of the assessment, is clear. The dynamic sampling strategy is intended as a complement to help with the latter. It does not mean dispense with the baseline data collection. Indeed the way NAWQA could become a more credible source of assessment information is if it could not only provide a spatially explicit benchmark of changes in water quality parameters, but also through dynamic sampling provide an explanation of the trends related to stressors and active management activities. As the NAWQA program moves forward with dynamic water quality monitoring, the committee urges NAWQA to evaluate trade-off's associated with and to achieve dynamic water quality monitoring. The committee hopes that the implementation of Cycle 3 will provide further clarity with respect to priorities and trade-offs. Also, in the final report the committee will answer the statement of task in its entirety and also speak to a number of issues and related topics raised during the review of this letter and deferred to the final report[2].

The NAWQA program has a history of working in the multidisciplinary interface, and this experience could benefit the USGS as it implements a re-alignment and in the face of certain fiscal

---

[2] Topics deferred to the final report include the history of NAWQA and what makes a national water quality assessment program, further probing priorities and trade-offs in light of current fiscal realities, a more detailed discussion of the technical aspects of the Science Plan, and a deeper discussion about internal collaborative approaches in light of the USGS reorganization.

realities. But again, the Cycle 3 Science Plan does not adequately describe how program goals are linked to not only the Water mission area but other mission areas in the realigned agency. NAWQA is a program of great value and strength to the USGS as the agency moves through this time of change, but the value and strength of NAWQA should be more fully articulated in the Science Plan. NAWQA should continue to seek collaborative opportunities within the agency and continue a common question and common goal oriented approach to collaboration.

Water availability, water use, and water quality will be among the nation's and the world's defining issues in the coming years. The interplay among water use, availability, and quality cannot be ignored: use affects quality and quality determines the availability of water for a particular use, including ecosystem use. The extent of water quality degradation from demographic and associated land use changes, agricultural chemicals, climate change, energy production, human use, and other factors must be characterized and quantified for effective water resources management. The NAWQA Program is looked to as a model for water quality monitoring outside the U.S. (Schindler, 2010). NAWQA is needed now more than ever to provide the scientific basis for wise management of water resources to provide clean water for humans and ecosystems and to strengthen the agency from within as it moves forward in this time of change.

Sincerely,

Donald I. Siegel, *Chair*
Committee to Review the USGS
National Water Quality Assessment
(NAWQA) Program

Attachment A: References
Attachment B: Committee Membership
Attachment C: Statement of Task
Attachment D: Acknowledgement of Reviewers
CC: S. Kimball & B. Werkheiser & D. Myers & S. Moulton & B. Wilber & G. Rowe

ATTACHMENT A

REFERENCES

Brakebill, J.W., S.W. Ator, and G.E. Schwarz. 2010. Sources of suspended-sediment flux in streams of the Chesapeake Bay Watershed: A regional application of the SPAROW model. *Journal of the American Water Resources Association* 46: 757-776.

Brigham, M. E., D. A. Wentz, G. R. Aiken, and D. P. Krabbenhoft. 2009. Mercury Cycling in Stream Ecosystems. 1. Water Column Chemistry and Transport. Environmental Science & Technology 43 (8):2720-2725.

Chasar, L. C., B. C. Scudder, A. R. Stewart, A. H. Bell, and G. R. Aiken. 2009. Mercury cycling in stream ecosystems -3. Trophic dynamics and methylmercury Bioaccumulation. *Environmental Science & Technology* 43 (8):2733-2739.

Kolpin, D. W., E. T. Furlong, M. T. Meyer, E. M. Thurman, S. D. Zaugg, L. B. Barber, and H. T. Buxton. 2002. Pharmaceuticals, hormones, and other organic wastewater contaminants in U.S. streams, 1999-2000—A national reconnaissance. Environmental Science and Technology 36 (6): 1202-1211.

Gartner, J. W. 2004. Estimating suspended solids concentrations from backscatter intensity measured by acoustic Doppler current profiler in San Francisco Bay, California. Marine Geology 211: 169-187.

Gartner, J. 2002. Estimation of suspended solids concentrations based on acoustic backscatter intensity: theoretical background. Proceedings of the Turbidity and Other Sediment Surrogates Workshop, April 30 – May 2, 2002, Reno, NV.

Gartner, J. W., R. T. Cheng, P. F. Wang, and K. Richter. 2001. Laboratory and field evaluations of the LISST-100 instrument for suspended particle size determinations. Marine Geology 175: 199-219.

Lipp, E. K., N. Schmidt, M. E. Luther, and J. B. Rose. 2001. Determining the Effects of El Niño–Southern Oscillation Events on Coastal Water Quality Estuaries 24(4): 491–497.

National Research Council (NRC). 1990. A Review of the U.S. Geological Survey National Water Quality Assessment Pilot Program. Washington, D.C.: National Academies Press.

National Research Council (NRC). 1992. Restoration of Aquatic Ecosystems. Washington, D.C.: National Academies Press.

National Research Council (NRC). 1995. Wetlands: Characteristics and Boundaries. Washington, D.C.: National Academies Press.

National Research Council (NRC). 2002. Opportunities to Improve the U.S. Geological Survey National Water Quality Assessment Program. Washington D.C.: National Academy Press.

National Research Council (NRC). 2009. Towards a Sustainable and Secure Water Future: A Leadership Role for the USGS. Washington, D.C.: National Academies Press.

National Research Council (NRC). 2010. Letter Report Assessing the USGS National Water Quality Assessment Program's Science Framework. Washington, D.C.: National Academies Press.

Pasquale, M. M. D., M. A. Lutz, M. E. Brigham, D. P. Krabbenhoft, G. R. Aiken, W. H. Orem, and B. D. Hall. 2009. Mercury cycling in stream ecosystems. 2. Benthic methylmercury production and bed sediment-pore water partitioning. Environmental Science & Technology 43(8):2726-2732.

Schindler, D. 2010. Tar Sands Need Solid Science. Nature 468: 499-501.

Thorne, P. D., and D. M. Hanes. 2002. A review of acoustic measurement of small-scale sediment processes. Continental Shelf Research 22(4): 603-632.

U.S. EPA. 2009. National Water Quality Inventory: Report to Congress, 2004 Reporting Cycle.

U.S. Geological Survey. 2007. Facing Tomorrow's Challenges—U.S. Geological Survey Science in the Decade 2007-2017. U.S. Geological Survey Circular 1309, 69 p. Available online at http://pubs.usgs.gov/circ/2007/1309/.
U.S. Geological Survey. 2010. The National Water-Quality Assessment Program—Science to Policy Management. June 30th, 2010. available at: http://water.usgs.gov/nawqa/xrel.pdf.

ATTACHMENT B

COMMITTEE ON PREPARING FOR THE THIRD DECADE (CYCLE 3) OF THE NATIONAL WATER-QUALITY ASSESSMENT PROGRAM

Donald I. Siegel, *Chair*, Syracuse University
Michael E. Campana, Oregon State University
Jennifer A. Field, Oregon State University
George R. Hallberg, The Cadmus Group, Inc.
Nancy K. Kim, State of New York Department of Health
Debra S. Knopman, RAND Corporation
Upmanu Lall, Columbia University
Walter R. Lynn, Cornell University
Judith L. Meyer, University of Georgia
David W. Schindler, NAS, University of Alberta
Deborah L. Swackhamer, University of Minnesota

NRC Staff
Laura J. Helsabeck, Study Director
Anita Hall, Project Assistant

ATTACHMENT C

STATEMENT OF TASK

The project will provide guidance to the U.S Geological Survey on the design and scope of the NAWQA program as it enters its third decade of water-quality assessments. The committee will assess accomplishments of the NAWQA program since its inception in 1991 by engaging in discussions with the Cycle 3 Planning Team, program scientists and managers, and external stakeholders and users of NAWQA data and scientific information. The committee will also review USGS internal reports on NAWQA's current design for monitoring, assessments, research, and relevance to key water topics. The main activities of the study committee will be to:

1. Provide guidance on the nature and priorities of current and future water-quality issues that will confront the nation over the next 10-15 years and address the following questions:
- Which issues are currently being addressed by NAWQA and how might the present design and associated assessments for addressing these issues be improved?
- Are there issues not currently being substantially addressed by NAWQA that should be considered for addition to the scope of NAWQA?

2. Provide advice on how NAWQA should approach these issues in Cycle 3 with respect to the following questions:
- What components of the Program—Surface Water Status and Trends; Ground-Water Status and Trends; Topical Understanding Studies; National Synthesis— should be retained or enhanced to better address national water-quality issues?
- What components of the program should change to improve how priority issues are addressed?
- Are there new Program components that should be added to NAWQA to enable the Program to better address and analyze National water-quality issues and related public policy issues?

3. Identify and assess opportunities for the NAWQA Program to better collaborate with other federal, state, and local government, non-governmental organizations, private industry, and academic stakeholders to assess the nation's current and emerging water quality issues.

4. Review strategic science and implementation plans for Cycle 3 for technical soundness and ability to meet stated objectives.

ATTACHMENT D

ACKNOWLEDGMENT OF REVIEWERS

This letter report has been reviewed in draft form by individuals chosen for their diverse perspectives and technical expertise, in accordance with procedures approved by the National Research Council's Report Review Committee. The purpose of this independent review is to provide candid and critical comments that will assist the institution in making its published report as sound as possible and to ensure that the report meets institutional standards for objectivity, evidence, and responsiveness to the study charge. The review comments and draft manuscript remain confidential to protect the integrity of the deliberative process.

We wish to thank the following individuals for their review of this report: Samuel N. Luoma, U.S. Geological Survey Emeritus; G. Tracy Mehan III, The Cadmus Group Inc.; Robert C. Ward, Colorado State University; Marylynn V. Yates, University of California, Riverside; and Jeanne M. VanBriesen, Carnegie Mellon.

Although the reviewers listed above have provided many constructive comments and suggestions, they were not asked to endorse the conclusions or recommendations nor did they see the final draft of the report before its release. The review of this report was overseen by Henry J. Vaux, Jr., University of California, Berkeley. Appointed by the National Research Council, he was responsible for making certain that an independent examination of this report was carried out in accordance with institutional procedures and that all review comments were carefully considered. Responsibility for the final content of this report rests entirely with the authoring committee and the institution.

Appendix C

Communicating Data and Information to Users

Communication of scientific activities is an accomplishment of the U.S. Geological Survey's (USGS's) National Water Quality Assessment (NAWQA) program, as discussed in Chapter 3. This appendix contains a more detailed accounting of the communication efforts to augment and support this accomplishment and further highlight the scope of these efforts.

The current budget for NAWQA communications effort is modest, approximately 1 to 2 percent of the total program budget. NAWQA communication activities are directed by the NAWQA Communications Coordinator. NAWQA has a contract with the Environmental and Energy Study Institute (EESI) to spearhead congressional briefings and liaison meetings. Separate communication resources are built into each of the program components for NAWQA, such as the surface water status and trends, groundwater status and trends, national synthesis programs, topical teams, and source-water assessments and are used to develop derivative products (such as the web, companion articles, fact sheets, video casts, etc.) that are associated with some of the relatively larger launches. NAWQA benefits from continued support from the USGS Office of Communication in developing companion products and social media tools. In public meetings of this committee, NAWQA leadership conveyed that the number of forums in which NAWQA results are presented continues to grow with the addition of more frequent and timely updating of web pages; improved mapping, querying, and access to related links on the web; and improved functionality of the data warehouse.

When disseminating NAWQA findings, the overall goal is to reach a broad audience of technical and non-technical consumers of NAWQA's information products and tools through multiple media, varying technical detail, and appealing graphics. Products are tailored to each targeted audience and structured around answering NAWQA's three core questions related to the status of the water resource, changes in quality and related conditions over time and space, and understanding of the factors influencing status and trends. The program's communication strategy follows a tiered approach, ranging from detailed scientific reports for technically trained audiences to one-page fact sheets for lay audiences. Thus, the program attempts to match the level of technical detail in its various information products to the needs of decision-makers and stakeholders who are trying to understand and make informed choices about policy, management, and investments related to water quality.

NAWQA typically employs an aggressive communications strategy in its release of a major report, which employs written products, briefings, and Internet-based formats. These reports are typically released with fact and briefing sheets, a press release, a dedicated web page linked to the main NAWQA program website, an email "blast" to stakeholders and decision-makers,[1] and stakeholder and congressional briefings. The dedicated webpage includes not only the main product but also a variety of other products including a Frequently Asked Questions document, downloadable graphics and tables with the raw and supporting data, and a podcast of key findings. The selection of products used depends on the scale and content of the report. The communication strategy is flexible and will expand or contract depending upon the attention received during release. It also ensures that NAWQA studies are communicated at both the local and national level. The release of the NAWQA pesticide circular in 2006 (Gilliom et al., 2006), which was a "major release," had a 10-part communication strategy and included a variety of derivative products and activities (Box C-1).

WRITTEN PRODUCTS

NAWQA's written publications are the foundation of the program's communications strategy. NAWQA publishes a variety of written publications that have no political agenda, target various audiences, and contain no specific policy recommendations. These publications come in a variety of formats targeting various audiences and subjects:

---

[1] NAWQA's Contacts Database includes approximately 1,700 relevant stakeholders who are prioritized according to their interest in the program judged by attendance at briefings, requests for NAWQA publications, etc. NAWQA uses this list and prioritization to distribute its publications and tailors outreach accordingly.

- *Circulars* are synthesis reports on a broad topic that are widely distributed, reaching a large audience. This includes scientists in government, industry, and academia; water managers; public-health officials; utilities; regulators; elected officials; and watershed groups and others in the general public.
- *Scientific Investigations Reports* contain information of lasting scientific interest because of significant data and interpretation.
- *Open File Reports* are publications that are released immediately and contain interpretive information such as supporting data referenced in another product or preliminary findings.
- *Water-Resources Information Reports* are a discontinued series of reports. When used, they contained hydrologic information of local interest.

BOX C-1
NAWQA Communication Strategy for Circular
1291: *Pesticides in the Nation's Streams and Ground Water, 1992-2001* **(released March, 2006)**

1. Conducted formal external review of the Circular (two formal USGS peer reviews, 15 external reviews);
2. Briefed federal agencies and others on major findings and implications (Department of the Interior [DOI], U.S. Environmental Protection Agency's [EPA's] Office of Water and Office of Pesticide Programs, U.S. Department of Agriculture [USDA];
3. Posted the Circular and companion materials on an internal website so USGS scientists (such as Water Science Center Directors) could become familiar with findings and implications;
4. Worked collaboratively with communications staff at EPA and USDA during release (exchanged press releases);
5. Briefed DOI external and public affairs stakeholders on key findings and communications plan;
6. Held congressional briefing;
7. Posted USGS Circular and companion materials on the NAWQA homepage;
8. Distributed Circular and companion materials to agencies and stakeholders (printed and electronic versions);
9. Derivative articles written to communicate information about specific issues of interest to selected organizations; and
10. Continued distribution through conferences, workshops, follow-up meetings, and briefings.

SOURCE: P. Hamilton, personal communication, June 25, 2010.

- *Fact Sheets* are abbreviated publications that summarize and provide details about various USGS activities, typically a Circular.
- *Techniques and Methods Reports* detail the techniques and methods used in NAWQA studies, both in the field and in the laboratory. These are useful and promote consistent water-quality monitoring among the scientific community at large.
- *Briefing sheets* are used as a tool at various meetings and on the website to brief interested parties. These do not appear in the NAWQA publications database and go through an unofficial peer review within NAWQA, in contrast to the more formal USGS peer-review process discussed below.
- *Journal articles* and *special journal issues* are often used as companions for the larger reports. For example, three papers on Mercury Cycling in Stream Ecosystems were published in the April 15, 2009, issue of *Environmental Science and Technology*, slightly prior to the report release in August of 2009.
- *Program design, strategy, and goal documents* are authored when the program experiences a shift in program design. These documents are often used to engage input from stakeholders such as the National Liaison Committee.

USGS publications go through the rigorous peer-review process. NAWQA supplements this by including an additional external reviewer(s), a practice started in 1991 with the Delmarva Circular in 1991 (Hamilton and Shedlock, 1992). For example, the pesticide circular released in 2006 had two formal USGS peer reviewers and 15 external reviewers including federal agencies, private and industry representatives, as well as representatives from trade, professional, and other non-profit organizations. When choosing this group of external reviewers, NAWQA's goal is to cover all relevant and potentially contrasting perspectives, i.e., non-regulatory, municipal, state, and tribal perspective (P. Hamilton, personal communication, June 17, 2010).

BRIEFINGS

NAWQA participates in approximately two congressional briefings a year, on a variety of topics that coincide with the release of new NAWQA research. Briefings typically employ a USGS scientists and a non-USGS scientist to address the topic. In large part, these briefings have been supported by the Water Environment Federation (a frequent co-sponsor) and EESI. Media and trade coverage is common and can include both local and national outlets. For example, the briefing on the NAWQA study assessing water-quality conditions of domestic wells across the United States held in March of 2009 had 42 media and trade press requests (P. Hamilton, NAWQA, personal communication, June 17, 2010).

When invited, NAWQA or USGS leadership will give congressional testimony on the program's scientific output. NAWQA hosted a Nitrogen and Phosphorus National Press Conference in 2007, was a press conference on SPAtially Referenced Regressions on Watershed Attributes (SPARROW) (P. Hamilton, personal communication, June 17, 2010).

NAWQA scientists frequently attend scientific meetings and present research on new monitoring and analytical methods, analysis of findings, and innovations in modeling and technology. Conferences include but are not limited to:

- America Benthological Society,
- Society of Environmental Toxicology and Chemistry,
- American Geophysical Union, American Geological Institute,
- American Water Resources Association—national and regional conferences,
- National Groundwater Association,
- Association of State Drinking Water Administrations,
- Geological Society of America, and
- American Association of Advancement of Science.

These presentations are important to overall success of the program because they increase the transparency of technical methods, and the credibility of the underlying analysis, and they engage a broader community of peers in the challenges of characterizing water quality over multiple time and spatial scales for a wide range of applications.

DATA AND OTHER MEDIA PRODUCTS ON THE INTERNET

During Cycle 2, the NAWQA program has expanded the use of digital media and appealing graphics to better communicate products and tools. NAWQA's primary web-based interface with the public is the program website,[2] has been significantly improved since the National Research Council encouraged NAWQA to improve it (NRC, 2002). Today, most users are satisfied or very satisfied with the program's website (Figure C-1).

The National Water Information System (NWIS)[3] is the USGS database, which houses streamflow, groundwater, and water-quality, and biology data. Ten to 15 years ago, NAWQA made its information accessible online through NWISweb, which is maintained by USGS's Office of Infor-

---

[2] See http://water.usgs.gov/nawqa.
[3] See http://waterdata.usgs.gov/nwis/qw.

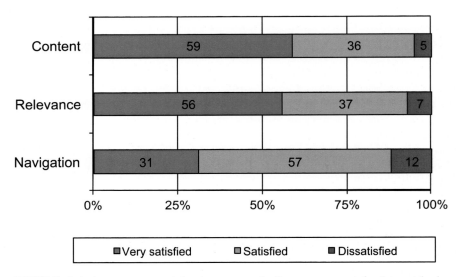

FIGURE C-1 A customer satisfaction survey indicates user satisfaction with the NAWQA website. SOURCE: USGS, personal communication.

mation. However, NAWQA's metadata were not completely compatible with NWISweb; thus, a subset of NWIS for NAWQA data, the NAWQA data warehouse, was born in 1999.[4] Containing information on approximately 2,000 water-quality and biological constituents, which are available for public use (Bell and Williamson, 2006), water-quality data are communicated to the public in a variety of formats including location maps, graphics, and links to NAWQA reports as well as instructions on data retrieval or exporting data. Biological information in the NAWQA data warehouse is not yet sophisticated, but part of this is a product of biological sampling constraints. USGS is currently building a database for biological data that will fit into NWISweb and the NAWQA data warehouse that will be released in late 2012, "BioData."[5] Even though this is a NAWQA-led effort, the biology database will serve and provide all biology data from water-related programs to the public (P. Hamilton, personal communication, May 13, 2009).

In recent years NAWQA has developed a social media presence by, for example, using USGS CoreCasts, an audio or video podcast, as a method of dissemination.[6] The first USGS CoreCast, on Hurricanes and Extreme

---

[4] See http://water.usgs.gov/nawqa/data.

[5] See https://aquatic.biodata.usgs.gov/.

[6] See http://www.usgs.gov/corecast/.

Storms, was released in August of 2007. Since then, NAWQA has developed and published several CoreCasts when a report or study is released. In February 2010, a CoreCast was released featuring the USGS NAWQA Transport of Anthropogenic and Natural Contaminants to Supply Wells or TANC effort.[7] The CoreCast explains the relevance of the study for the educated lay person, notes that the results of the study illustrate why some public-supply wells are more vulnerable to contaminants in aquifers than others, and mentions that the study provides information for public supply well managers to protect their drinking water supply—conclusions that are very important to the public. The CoreCast continues by linking the video to USGS fact sheets providing the viewer with a mechanism for obtaining more information.

For the recent study on the Effects of Urbanization on Stream Ecosystems, NAWQA scientists developed a complimentary set of video casts, which received 7,000 page views the day the study was released (G. McMahon, personal communication, June 21, 2010). The urbanization story is reaching the international community despite the geographical focus on the United States; upon release, Spanish Univision reported on the study, and the NAWQA leadership is considering using Spanish subtitles in its CoreCasts (P. Hamilton, personal communication, June 17, 2010).

USGS, through the Office of Communications, is also participating in additional social media outlets including Twitter, Facebook, and YouTube. NAWQA is contemplating involvement in these outlets. The USGS Office of Communications commented on NAWQA's 2009 mercury study upon release and received 20,000 "tweets" in response and discussion. The public took a particular interest in understanding if fish were safe to eat. However, it is reasonable to note that the public might not show the same interest in carbonate aquifers or NAWQA's more technically oriented products.

REFERENCES

Bell, R. W., and A. K. Williamson. 2006. Data Delivery and Mapping Over the Web: National Water-Quality Assessment Data Warehouse. USGS Fact Sheet 2006-3101.

Gilliom, R. J., J. E. Barbash, C. G. Crawford, P. A. Hamilton, J. D. Martin, N. Nakagaki, L. H. Nowell, J. C. Scott, P. E. Stackelberg, G. P. Thelin, and D. M. Wolock. 2006. Pesticides in the Nation's Streams and Ground Water, 1992-2001. U.S. Geological Survey Circular 1291.

Hamilton, P. A., and R. J. Shedlock. 1992. Are fertilizers and pesticides in the ground water—A case study of the Delmarva Peninsula, Delaware, Maryland, and Virginia: U.S. Geological Survey Circular 1080. 16 pp.

NRC (National Research Council). 2002. Opportunities to Improve the U.S. Geological Survey National Water Quality Assessment Program. Washington, DC: National Academies Press.

---

[7] See http://oh.water.usgs.gov/tanc/NAWQATANC.htm.

Appendix D

Contributors to the Report, *Preparing for the Third Decade of the National Water-Quality Assessment Program*

During the course of this study, numerous persons contributed to the development of this report. Some provided material and talks at the request of the committee, some provided unsolicited material, and others provided advice for the committee to consider. The committee would like to thank all of these persons for their interest and support for this effort.

Joseph D. Ayotte, Chief, Groundwater Investigations and Research Section, New Hampshire–Vermont Water Science Center, USGS
Joe Beaman, U.S. Environmental Protection Agency
Betsy Behl, U.S. Environmental Protection Agency
Ken Belitz, Project Chief, California Ground-Water Ambient Monitoring Assessment, USGS
Nate Booth, IT Specialist, Center for Integrated Data Analytics, USGS
Sally Brady, Staff Scientist
Herb Buxton, Program Coordinator, Toxics Substances Hydrology Program, USGS
Judy Campbell Byrd, Environmental and Energy Study Institute
Tom Carpenter, U.S. Environmental Protection Agency
Charles Demas, Director, Louisiana Water Science Center, USGS
Neil Dubrovsky, Chief, NAWQA Nutrients National Synthesis, USGS
Hedeff Essaid, Research Hydrologist, Menlo Park, CA, USGS
Robert Gilliom, Chief, Pesticide National Synthesis Team, USGS
Paul Gruber, National Ground Water Association

Pixie Hamilton, NAWQA Communications Coordinator (*through summer, 2011*), Cooperative Water Program National Coordinator, USGS
Roger Helm, U.S. Fish and Wildlife Service
Robert M. Hirsch, Research Hydrologist, USGS
Susan Holdsworth, U.S. Environmental Protection Agency
Anne Hoos, Hydrologist, Tennessee Water Science Center, USGS
David Kidwell, National Oceanic and Atmospheric Administration
Matt Larsen, Associate Director, Climate and Land Use Change, USGS
Dennis Lynch, Director, Oregon Water Science Center, USGS
Mark Munn, Ecologist, Washington Water Science Center, USGS
Donna Myers, Chief, Office of Water Quality, USGS
Robin O'Malley, The H. John Heinz III Center for Science, Economics, and the Environment
Darrell Osterhoudt, Association of State Drinking Water Administrators
Tim Parker, National Ground Water Association
Marc Ribaudo, U.S. Department of Agriculture
Gary Rowe, Regional NAWQA Program Coordinator and Head of the NAWQA Cycle 3 Planning Team, USGS
Nancy Stoner, Natural Resources Defense Council (*through February 2010*)
Ione Taylor, Associate Director, Energy, Minerals, and Environmental Health, USGS
Joanne Taylor, Communications Specialist for the Director, USGS
Joshua F. Valder, USGS
A. B. Wade, Public Affairs Officer, USGS
Barbara Wainman, Director of the USGS Office of Communications, USGS
William Werkheiser, Associate Director, Water, USGS
Bill Wilbur, Chief, National Water-Quality Assessment Program, USGS
Dave Wolock, Lead Scientist, NAWQA Hydrologic Systems Team, USGS
Paul Young, Deputy Associate Director, Energy, Minerals, and Environmental Health, USGS
John S. Zogorski, Chief, Volatile Organic Compounds National Synthesis Team, USGS

Appendix E

Biographical Information: Committee on Preparing for the Third Decade (Cycle 3) of the National Water-Quality Assessment (NAWQA) Program

Donald I. Siegel is a professor of geology at Syracuse University, where he teaches graduate courses in hydrogeology and aqueous geochemistry. He holds B.S. and M.S. degrees in geology from the University of Rhode Island and Pennsylvania State University, respectively, and a Ph.D. in hydrogeology from the University of Minnesota. His research interests are in solute transport at both local and regional scales, wetland-groundwater interaction, and paleohydrogeology. He has been a member of several National Research Council (NRC) committees including the Water Science and Technology Board's (WSTB's) Committee on Water Resources Activities at the U.S. Geological Survey (USGS). Dr. Siegel is also a member of the WSTB and served as Chair of the WSTB Committee on River Science at the U.S. Geological Survey (*River Science at the U.S. Geological Survey*, 2007).

Michael E. Campana is a professor of Hydrogeology and Water Resources in the College of Earth, Ocean, and Atmospheric Sciences at Oregon State University (OSU) and former Director of its Institute for Water and Watersheds. Prior to joining OSU in 2006 he held the Albert J. and Mary Jane Black Chair of Hydrogeology, directed the Water Resources Program at the University of New Mexico, was a research hydrologist at the Desert Research Institute, and taught in the University of Nevada-Reno's Hydrologic Sciences Program. He has supervised 68 graduate students. His research and interests include hydrophilanthropy, water resources management and policy, communications, transboundary water resources, regional hydrogeology, and surface water-groundwater interactions. He was a Fulbright Scholar to Belize and a Visiting Scientist at Research Institute for Groundwater (Egypt) and the IAEA in Vienna. Central America and the South Caucasus are the current foci of his international work. He has served on six National Research Council committees. Dr. Campana is founder, president, and treasurer of the Ann Campana Judge Foundation (www. acjfoundation.org), a 501(c)(3) charitable foundation that funds and undertakes projects related to water, sanitation, and hygiene (WaSH) in Central America. He operates the WaterWired blog and Twitter. He earned a B.S. in geology from the College of William and Mary and M.S. and Ph.D. degrees in hydrology from the University of Arizona.

Jennifer A. Field is a professor at Oregon State University in the Department of Environmental and Molecular Toxicology. Dr. Field holds a B.S. in earth science from Northland College and a Ph.D. from Colorado School of Mines. An analytical chemist, her research interest is in understanding the occurrence, transport, and fate of contaminants in groundwater, surface water, and waste water effluent. An expert in analytical methods development, she currently focuses on detection of organic contaminants such as illicit drugs, fluorine-containing compounds, and nanomaterials in water. She held a student appointment with the USGS Water Resource Discipline National Research Program from 1987-1990. Dr. Field is an associate editor for *Environmental Science &Technology*.

George R. Hallberg is a principal with the Cadmus Group, Inc., in Watertown, Massachusetts, conducting environmental science and policy research, regulatory analysis, and management services. Previously he was associate director and chief of environmental research at the University of Iowa's environmental and public health laboratory and at the Iowa Department of Natural Resources. Dr. Hallberg was also an adjunct professor at both the University of Iowa and Iowa State University. He chaired the NRC Committee on Water Resources Activities at the U.S. Geological Survey (2009) and the Committee on Opportunities to Improve the USGS National Water-Quality Assessment (NAWQA) program (2002); he also has served on the Committee for Assessment of Water Resources Research, and others, and as a member of the NRC Board on Agriculture and Natural Resources. He served on the U.S. Environmental Protection Agency (EPA) National Advisory Council for Environmental Policy and Technology and on the Office of Water's Management Advisory Group. He is a National Associate of The National Academies. His research interests include environmental monitoring and assessment, agricultural-environmental impacts, chemical and nutrient fate and transport, contaminant occurrence and trends in drinking water, and health effects of environmental contaminants. Dr. Hallberg received a B.A. in geology from Augustana College and a Ph.D. in geology from the University of Iowa.

Nancy K. Kim is senior executive for the Center for Environmental Health, New York State Department of Health. Dr. Kim is also associate professor, School of Public Health, State University at Albany. She was director, Bureau of Toxic Substance Assessment, New York State Department of Health; director, Division of Environmental Health Assessment (Bureau of Toxic Substance Assessment and Bureau of Environmental Exposure Investigation); and director of the Center for Environmental Health, New York State Department of Health. Dr. Kim's expertise is in toxicological evaluations, exposure assessments, risk assessments, structural activity correlations, and quantitative relationships between toxicological parameters. Dr. Kim received her B.A. in chemistry from the University of Delaware and her M.S. and Ph.D. in chemistry from Northwestern University. She has served on four prior NRC committees.

Debra S. Knopman is vice president and director of the RAND Corporation Infrastructure, Safety and Environment (ISE) division. ISE leads RAND's policy research on homeland security, safety and justice, environment, energy, and economic development, and transportation, space and technology. Her expertise includes energy, the environment, water resources, and public administration. Dr. Knopman served from 1997 to 2003 as a member of the U.S. Nuclear Waste Technical Review Board, and she chaired the Board's Site Characterization Panel. She was previously the director of the Center for Innovation and the Environment at the Progressive Policy Institute, the deputy assistant Secretary for water and science at the Department of the Interior, a hydrologist at the U.S. Geological Survey, and a staff member for the U.S. Senate Environment and Public Works Committee. She has also served on several NRC committees and was a member of the Space Studies Board.

Upmanu Lall is professor and chair of Earth and Environmental Engineering at Columbia University. His principal areas of expertise are statistical and numerical modeling of hydrologic and climatic systems and water resource systems planning and management. Dr. Lall has more than 25 years of experience as a hydrologist. He has been the principal investigator on a number of research projects funded by the U.S. Geological Survey, the National Science Foundation (NSF), the U.S. Air Force, the National Oceanic and Atmospheric Administration (NOAA), U.S. Bureau of Reclamation, the Department of Energy, and Utah and Florida state agencies. These projects have covered water quantity and quality and energy resource management, flood analysis, groundwater modeling and subsurface characterization, climate modeling, and the development of statistical and mathematical modeling methods. He has been involved as a consultant with specialization in groundwater flow and contaminant transport modeling covering mining operations, streamflow modeling and water balance, risk and environmental impact assessment, site hydrologic evaluation, and as a reviewer and as an expert on a number of other hydrologic problems. He has also taught more than 20 distinct university courses. Dr. Lall has served on two prior NRC committees.

Walter R. Lynn became a member of the Civil and Environmental Engineering Faculty in 1961 at Cornell University. He directed the Program on Science, Technology, and Society for 8 years, and he served as director of Cornell's Center for Environmental Research and director of the School of Civil and Environmental Engineering. He also served as faculty trustee from 1980-1985 and as a member of the Board of Directors of the Cornell Research Foundation. During his term as dean of the university faculty, he recognized several issues that faced the faculty, including the quality of undergraduate education and the status of federal support for research, among others. He was instrumental in the establishment of the Weiss Teaching Awards recognizing outstanding undergraduate teaching faculty. Dr. Lynn was named emeritus on February 1, 1998, and served as university ombudsman afterward. Dr. Lynn was part of the National Research Council report on the 1984 NAWQA proposal that originated the program. He served on numerous NRC committees as a participant and chair as well as a WSTB member and chair. Dr. Lynn received the USDI Conservation Service Award in 1994 and a USGS Field Office Dedication in 1999. In 2003 Dr. Lynn became a National Associate of the National Academy of Sciences.

Judith L. Meyer is distinguished fellow at the River Basin Center and professor emeritus at Odum School of Ecology, University of Georgia. Dr. Meyer has served on the Water Science and Technology Board, the Board on Environmental Studies and Toxicology, and several NRC committees. She is a past president of the Ecological Society of America. She currently serves on EPA's Science Advisory Board and on the Scientific and Technical Advisory Committee of American Rivers. She received the award of Excellence in Benthic Science from the North American Benthological Society and the Naumann-Thienemann medal from the International Limnological Society. Dr. Meyer was named a Clean Water Act Hero by Clean Water Network and is a AAAS fellow. Her expertise is in river and stream ecosystems with emphasis on nutrient dynamics, microbial food webs, riparian zones, ecosystem management, river restoration, and urban rivers. Dr. Meyer received a Ph.D. in 1978 from Cornell University. She was part of the 1990 and 2002 NRC NAWQA reports.

David W. Schindler (NAS) is Killam Memorial Professor of Ecology at the University of Alberta, Edmonton. From 1968 to 1989, he founded and directed the Experimental Lakes Project of the Canadian Department of Fisheries and Oceans near Kenora, Ontario, conducting interdisciplinary research on the effects of eutrophication, acid rain, radioactive elements, and climate change on boreal ecosystems. Dr. Schindler is the world leader in understanding lake biogeochemistry. His pioneering studies involving whole lake experiments convincingly verified the phosphorus-eutrophication connection and the impact of atmospheric acidification on lake production. He has brilliantly revealed the effects of ultraviolet (UV) radiation and airborne organochlorine contaminants on boreal lakes. His current research interests include the study of fisheries management in mountain lakes, the biomagnification of organochlorines in food chains, effects of climate change and UV radiation on lakes, and global carbon and nitrogen budgets. He received a B.S. in 1962 from North Dakota State University and a Ph.D. in 1966 from Oxford University. Dr. Schindler is a member of the National Academy of Sciences and is an active NRC committee participant.

Deborah L. Swackhamer is professor of environmental chemistry in the Division of Environmental Health Sciences in the School of Public Health at the University of Minnesota, the Charles Denny Chair of Science, Technology and Public Policy in the Humphrey School of Public Affairs, and co-director of the University's Water Resources Center. She received a B.A. in chemistry from Grinnell College (Grinnell, Iowa) and an M.S. and Ph.D. from the University of Wisconsin-Madison in water chemistry and limnology, and oceanography, respectively. Dr. Swackhamer currently serves as chair of the Science Advisory Board of the U.S. Environmental Protection Agency, and on the Science Advisory Board of the International Joint Commission of the United States and Canada. She serves on the Minnesota Clean Water Council. Dr. Swackhamer is a member of the editorial advisory board for the journal *Environmental Science & Technology*, and she chairs the editorial advisory board of the *Journal of Environmental Monitoring*. She is a fellow in the Royal Society of Chemistry in the United Kingdom. Dr. Swackhamer received the Harvey G. Rogers Award from the Minnesota Public Health Association in June 2007, the 2009 Founders Award from the Society of Environmental Toxicology and Chemistry, and the 2010 Ada Comstock Award from the University of Minnesota. She has served on three prior NRC committees.